Thomas Staub

Development, Testing, Deployment & Operation of Wireless Mesh Networks

Thomas Staub

Development, Testing, Deployment & Operation of Wireless Mesh Networks

Addressing various challenges encountered in the life cycle of Wireless Mesh Networks

Südwestdeutscher Verlag für Hochschulschriften

Impressum/Imprint (nur für Deutschland/only for Germany)
Bibliografische Information der Deutschen Nationalbibliothek: Die Deutsche Nationalbibliothek verzeichnet diese Publikation in der Deutschen Nationalbibliografie; detaillierte bibliografische Daten sind im Internet über http://dnb.d-nb.de abrufbar.
Alle in diesem Buch genannten Marken und Produktnamen unterliegen warenzeichen-, marken- oder patentrechtlichem Schutz bzw. sind Warenzeichen oder eingetragene Warenzeichen der jeweiligen Inhaber. Die Wiedergabe von Marken, Produktnamen, Gebrauchsnamen, Handelsnamen, Warenbezeichnungen u.s.w. in diesem Werk berechtigt auch ohne besondere Kennzeichnung nicht zu der Annahme, dass solche Namen im Sinne der Warenzeichen- und Markenschutzgesetzgebung als frei zu betrachten wären und daher von jedermann benutzt werden dürften.

Coverbild: www.ingimage.com

Verlag: Südwestdeutscher Verlag für Hochschulschriften GmbH & Co. KG
Heinrich-Böcking-Str. 6-8, 66121 Saarbrücken, Deutschland
Telefon +49 681 37 20 271-1, Telefax +49 681 37 20 271-0
Email: info@svh-verlag.de

Approved by: Bern, Universität, Diss., 2011

Herstellung in Deutschland (siehe letzte Seite)
ISBN: 978-3-8381-3358-4

Imprint (only for USA, GB)
Bibliographic information published by the Deutsche Nationalbibliothek: The Deutsche Nationalbibliothek lists this publication in the Deutsche Nationalbibliografie; detailed bibliographic data are available in the Internet at http://dnb.d-nb.de.
Any brand names and product names mentioned in this book are subject to trademark, brand or patent protection and are trademarks or registered trademarks of their respective holders. The use of brand names, product names, common names, trade names, product descriptions etc. even without a particular marking in this works is in no way to be construed to mean that such names may be regarded as unrestricted in respect of trademark and brand protection legislation and could thus be used by anyone.

Cover image: www.ingimage.com

Publisher: Südwestdeutscher Verlag für Hochschulschriften GmbH & Co. KG
Heinrich-Böcking-Str. 6-8, 66121 Saarbrücken, Germany
Phone +49 681 37 20 271-1, Fax +49 681 37 20 271-0
Email: info@svh-verlag.de

Printed in the U.S.A.
Printed in the U.K. by (see last page)
ISBN: 978-3-8381-3358-4

Copyright © 2012 by the author and Südwestdeutscher Verlag für Hochschulschriften GmbH & Co. KG and licensors
All rights reserved. Saarbrücken 2012

Abstract

Wireless Mesh Networks (WMNs) are a key technology to provide inexpensive ubiquitous network access to end users and sensing equipment in urban, rural, and developing areas. WMNs seamlessly integrate with existing traditional fixed or cellular networks and extend their network coverage. Like any other network, a WMN and the services running on top of it go through a life cycle consisting of *development, testing, deployment* and *operation*. This thesis contributes solutions addressing various challenges encountered in each phase of the life cycle.

First, we designed a flexible framework for development and testing of new protocols and architectures. The framework is based on traffic interception by a virtual wireless interface and on network emulation. It offers instruments to comprehensively test real prototype implementations within a well controllable environment. Hence, diverse conditions and scenarios can be efficiently evaluated.

Second, a build system for an own embedded Linux distribution tailored to mesh nodes has been implemented. It supports cross-compilation for various node platforms and incorporates features of a novel management architecture. We designed this management architecture to safely handle configuration and software updates and to avoid costly on-site repairs in WMNs. The architecture also guarantees the remote accessibility to all network nodes in the presence of configuration errors and faulty updates.

Further, we tested the applicability of WMNs for environmental monitoring in an outdoor deployment. The deployment delivered expert knowledge concerning the deployment processes, such as identification of best practises, evaluating the equipment and its usability for future outdoor deployments.

As another practical application of WMNs, a video conferencing system on construction sites was investigated. The temporally deployed WMN in such scenario was designed to be easily set up and operated by a non-expert user.

Finally, a prototype implementation has proven the feasibility of an autonomous deployment of a WMN using unmanned aerial vehicles for communication in emergency and disaster scenarios.

Contents

List of Figures   v

List of Tables   x

1 Introduction   **1**
    1.1 Problem Statement 3
        1.1.1 Development Phase 3
        1.1.2 Testing/Implementation Phase 4
        1.1.3 Deployment Phase 5
        1.1.4 Network Operation Phase 5
    1.2 Research Contributions 6
        1.2.1 Operating System and Management for WMNs 7
        1.2.2 Development and Testing Support 8
        1.2.3 WMN for Environmental Monitoring 8
        1.2.4 Deployment Support for an Ad-Hoc WMN 9
        1.2.5 Autonomous Deployment of a WMN using Unmanned Aerial Vehicles 9
    1.3 Summary of Contributions 10
    1.4 Thesis Outline 11

2 Related Work   **13**
    2.1 Wireless Mesh Networks 13
        2.1.1 Routing 16
        2.1.2 Multi-Channel Communication 21
        2.1.3 Network Management 23
    2.2 WMN Nodes 25
        2.2.1 WMN Hardware Platforms 25
        2.2.2 Embedded Operating Systems Distributions 28
    2.3 Network Simulation and Emulation 29
        2.3.1 Network Simulation 30
        2.3.2 Network Emulation 32

2.4	Existing WMN Deployments and Testbeds	35
	2.4.1 Outdoor Deployments	35
	2.4.2 Testbeds	36
2.5	Deployment Support for Wireless Mesh Networks	38
2.6	Unmanned Aerial Vehicle Hardware	39
2.7	Regulations	40
2.8	Conclusions	41

I General Frameworks and Tools   43

3 Operating System and Management for WMNs   45

3.1	Introduction	46
3.2	ADAM: Concept and Architecture	48
	3.2.1 Decentralised Distribution Mechanism	48
	3.2.2 Self-Healing	49
	3.2.3 Separation of Software and Configuration Data	50
3.3	ADAM: Build System	51
3.4	ADAM: Management Operation	55
	3.4.1 ADAM Distribution Engine	56
	3.4.2 Configuration Module	57
	3.4.3 New Node Module	57
	3.4.4 Software Update Module	59
	3.4.5 Command Module	61
	3.4.6 Lost Node Detection	61
	3.4.7 Web-Based Management	62
3.5	Evaluation	65
3.6	Conclusions	67

4 Development and Testing Support   69

4.1	Introduction	70
4.2	VirtualMesh Concept and Architecture	71
4.3	VirtualMesh Communication Protocol	74
4.4	Host Virtualisation	77
4.5	Client Implementation	77
4.6	Wireless Simulation Server	84
	4.6.1 Components	84
	4.6.2 Message Flow	86
	4.6.3 Protocols	87
4.7	ADAM/VirtualMesh Integration	92
4.8	Evaluation	93

	4.8.1 VirtualMesh Test Setup	94
	4.8.2 Functional Evaluation using ADAM	96
	4.8.3 Performance Evaluation	98
4.9	Conclusions	106

II Application Specific Use Cases — 111

5 WMN for Environmental Monitoring — 113
- 5.1 Introduction . . . 114
- 5.2 Application Scenario . . . 114
- 5.3 Equipment . . . 116
  - 5.3.1 Mesh Nodes and Antennas . . . 116
  - 5.3.2 Power Supply for the Mesh Nodes . . . 118
  - 5.3.3 Masts . . . 119
  - 5.3.4 Wall Mounting . . . 120
  - 5.3.5 Tools and Utilities . . . 120
- 5.4 Deployment Parameters . . . 121
- 5.5 Software . . . 123
- 5.6 Planning, Predeployment, and Deployment . . . 124
- 5.7 Evaluation . . . 130
- 5.8 Conclusions . . . 134

6 Deployment Support for an Ad-Hoc WMN — 137
- 6.1 Motivation . . . 138
- 6.2 OViS Concepts and Architecture . . . 140
  - 6.2.1 Requirements . . . 141
  - 6.2.2 Network Setup . . . 142
  - 6.2.3 Multi-Channel Communication . . . 143
  - 6.2.4 Message Flow between OViS Client and the Mesh Node . . . 144
- 6.3 OViS Mesh Nodes . . . 147
- 6.4 OViS Deployment Applications . . . 149
- 6.5 Evaluation . . . 157
  - 6.5.1 Determination of RSSI Thresholds . . . 157
  - 6.5.2 OViS Performance Evaluation . . . 158
  - 6.5.3 Multi-Hop Throughput . . . 160
  - 6.5.4 Multi-Channel Performance . . . 161
- 6.6 Conclusions . . . 162

7 Autonomous Deployment of a Wireless Mesh Network using Unmanned Aerial Vehicles   **165**
  7.1 Introduction 166
  7.2 Scenario 167
      7.2.1 Search Mode 167
      7.2.2 Positioning of UAVs 168
      7.2.3 Single Airborne Relay 168
      7.2.4 Multi-Hop Airborne Relay 170
  7.3 System Components 171
      7.3.1 Communication Types 172
      7.3.2 Prototype 172
  7.4 Communication Protocol 175
      7.4.1 Protocol Messages 175
      7.4.2 Message Flow 176
  7.5 Remote Control Client 179
  7.6 Evaluation 181
      7.6.1 Determination of Optimal Signal Strength Thresholds 181
      7.6.2 Multi-Hop Performance 183
      7.6.3 Effect of Too Far Away Nodes 184
  7.7 Conclusions 184

8 Conclusions and Outlook   **187**
  8.1 Summary 189
  8.2 Outlook 190

9 Acronyms   **193**

Bibliography   **197**

List of Figures

1.1 Wireless mesh network. 1
1.2 Usual life cycle of a network consisting of development, testing / implementation, deployment, and operation. 2
1.3 Systematic overview of contributions. 6

2.1 Hybrid wireless mesh network. 14
2.2 Optimised flooding OLSR: classical vs. multi-point-relay flooding. . . 18
    (a)   Classical full flooding. 18
    (b)   Multipoint Relays. 18
2.3 PCEngines WRAP.2C with an indoor case. 26
2.4 PC Engines ALIX.3d2 system board. 27
2.5 Meraki Mini with an indoor case. 27
2.6 OpenMesh professional router OM1P. 28
2.7 Quadrocopter from HiSystems Ltd. - a small unmanned aerial vehicle. 39
2.8 Quadrocopter flight electronics: main processor board Flight-Ctrl, four brush-less controllers, GPS module, NaviCtrl with three-axis magnetic field sensor. 40

3.1 Example of a WMN: One node is temporarily unavailable, e.g., due to lack of power. Another node is added to the network for the first time. Multiple nodes can provide management functionality for the network. 47
3.2 Distribution of node configuration and software updates. 49
    (a)   Nodes periodically check for updates (green arrows). A new configuration is injected at a management node (M) or a normal node. 49
    (b)   First nodes (A, B) get the update from node M (orange arrows). 49
    (c)   Next nodes (C, D) get the update from node A and B. 49
3.3 ADAM: Steps of the build and set-up process for a node. 52
3.4 Run time layout of system RAM and the secondary storage for PCEngines ALIX/WRAP, Meraki Mini and OpenMesh OM1P nodes. 54

3.5	Detailed boot process.	54
3.6	General ADAM management architecture.	55
3.7	Integration of a new node into an existing network.	58
	(a) New node searches for networks having an ESSID that matches an IPv6 prefix.	58
	(b) New node automatically configures a valid IPv6 address and tries to get its configuration from neighbours. After the new node has received its configuration, it is fully integrated into the network.	58
	(c) If no configuration is available, the node announces its state to a management node. The user has to generate a new configuration. The new node is integrated in the network after having received this configuration.	58
	(d) After the new node has received its configuration, it is fully integrated into the network.	58
3.8	Safe software update process for Linux kernel and root file system with automatic fall back to previous software image.	60
3.9	ADAM: Management of network configuration.	62
3.10	ADAM: Modification of selected network configuration.	63
3.11	ADAM: Edit the network configuration of an individual node.	64
4.1	General concept: Traffic interception and emulation of the wireless medium via subdivision of the network stack.	72
4.2	VirtualMesh architecture with real nodes, virtualised nodes, and the simulation model.	73
4.3	Message format to communicate with the model: data transmission and node registration.	75
4.4	A node with native Linux network stack (a) and a node with our virtual network interface (b) (*iwconnect*) communicating with the OMNeT++ simulation model.	80
	(a) Physical interface.	80
	(b) Virtual interface	80
4.5	Access to virtual interfaces and its parameters using *libvif*.	82
4.6	Packet flow between two nodes interconnected by the OMNeT++ simulation model.	83
4.7	Message flow inside the simulation model 'WlanModel'.	86
4.8	Node registration.	88
4.9	Node de-registration.	89
4.10	Node configuration.	92

4.11 Experimental setup with multiple virtualised wireless nodes running on a XEN virtualisation server and a simulation server hosting the WlanModel. 95
4.12 Summarised RTT results for quantifying infrastructure network delay. 99
4.13 *iwconnect*/VirtualMesh communication protocol RTT overhead with respect to payload size. 100
4.14 *iwconnect*/emulation protocol RTT overhead with respect to transmission interval. 101
4.15 RTT with various payload sizes (distance = 300m, transmission interval = 1s). 102
4.16 RTT with concurrent streams (distance = 300m, transmission interval = 1s). 102
4.17 *WlanModel* scalability (distance = 300m, transmission interval = 1s, payload size = 56B). 103
4.18 *WlanModel* multi-hop behaviour (distance = 500m, transmission interval = 1s, payload size = 56B). 104
4.19 Aggregated throughput for parallel transfers using TCP and UDP. . . 105
4.20 Multi-hop throughput results. 106

5.1 Map of Switzerland with the location of CTI-Mesh network. 115
5.2 CTI-Mesh network deployed in the area Neuchâtel - Payerne, Switzerland. 115
5.3 Deployed nodes. 116
  (a) Node03 in Corges. 116
  (b) Node04 in Belmont. 116
5.4 Node05 deployed near Belmont. 117
5.5 Node06 on the platform roof of the MeteoSwiss building in Payerne. . 117
5.6 Water protected enclosure. 118
5.7 Power supply box with solar charger, lead acid battery, passive PoE adapter, yellow electric cable, and twisted pair cable. 119
5.8 Assembling *node01* on the roof of the University of Neuchâtel. 120
5.9 Helpful special tools. 121
  (a) Amplitude compass. 121
  (b) Mast level. 121
  (c) Socket wrench with ratchet handle. 121
5.10 Complete assembly of telescopic mast in horizontal position before final setup. 126
5.11 Concrete paving slab to prevent sinking in of the tripod, sand bag and iron stake to stabilise mast. 127
5.12 Primarily used fixedly mounted threepart guying clamp and its replacement part, a movable guying clamp to prevent torsion of mast. . 128

5.13 Broken mast due to strong winds and loose guying (*node02*). 129
5.14 Screenshot of IP camera streaming over WMN. 130
5.15 TCP bandwidth for the connections to *node01*. 131
5.16 TCP bandwidth for each link. 132
5.17 ETX values for best route from *node01* to *node06*. 133
5.18 Route availablity to *node06*/IP camera at *node01*. 133
5.19 Received signal strengths for all six links. 134

6.1 Motivation for OViS: An electrician requires instructions to solve an issue at the switching unit in the basement of a building. Unfortunately, there is no reception of cellular networks in the basement. . . 138
6.2 OViS: A temporary WMN provides Internet connectivity in the basement and, therefore, enables video-conferencing to discuss problems comfortably and efficiently. 139
6.3 Stepwise deployment of the temporary network for OViS. 140
 (a) Gateway node deployed. 140
 (b) Intermediate node deployed. 140
 (c) Complete network deployed. 140
6.4 OViS network topology: Full OLSR (IPv4). 142
6.5 OViS network topology: OLSR (IPv6) with IPv4-in-IPv6 tunnel. . . . 142
6.6 OViS network topology: OLSR (IPv4). 143
6.7 OViS: Deployment and configuration steps for multi-channel communication. 145
 (a) Step 1 145
 (b) Step 2 145
 (c) Step 3 145
6.8 OViS: Message sequence for the deployment of a mesh node. 146
6.9 Prototype of a battery-powered OViS WMN node. 147
6.10 OViS components on a wireless mesh node. 148
6.11 OViS Support Process: Deployment of a temporary communication infrastructure for on-site video conferencing by an inexperienced user. 149
6.12 Command-line and graphical OViS clients for personl computers. . . 151
 (a) OViS command-line client 151
 (b) Mac OS X OViS client 151
6.13 OViS full-screen *kiosk* application optimised for the Asus R2H UMPC.152
6.14 OViS deployment process guided by an Android application (Part I). 153
6.15 OViS deployment process guided by an Android application (Part II). 154
6.16 OViS deployment application for iOS (running on iPhone). 155
6.17 OViS deployment application for iOS (running on iPad). 156
6.18 Achievable single hop throughput in relation with the received signal strength indicator (RSSI). 157

6.19 OviS deployment test scenario. 158
6.20 Throughput of different deployments: non-guided deployment, OviS deployment, and manually optimised. 159
6.21 Signal strengths achieved after deployment with OViS and after manual optimisation with relocating the nodes and aligning the antennas. 160
6.22 Throughput depending on the number of hops. 161
6.23 Throughput in a two hop scenario using single and multi-channel communication. 162

7.1 Flying UAV swarm carrying a temporary WMN. 167
7.2 Multi-Hop Airborne Relay Scenario. 168
7.3 Process of connecting two distant clients by one single flying WMN node (airborne relay). 169
7.4 System components of UAVNet: WMN node with UAV controller and IEEE 802.11s mesh access point (MAP), UAV electronics and UAV client. 171
7.5 Communication types in UAVNet: Serial to interconnect WMN node and UAV, IEEE 802.11s for data and control traffic. 173
7.6 UAVNet: Flight electronics connected by a serial connection to the WMN node. 174
7.7 Flying quadrocopter UAV carrying a WMN node. 174
7.8 UAVNet: Protocol messages. 175
7.9 Message flow for a scenario with and without known direction towards the location of the second client (manual and autonomous search). . . 177
7.10 Message flow for subscribing to notification service. 178
7.11 Remote Control Application (iPad): Selection of network deployment scenario. 179
7.12 Remote Control Application (iPhone): setting the scenario, confirmation, and monitoring of flying UAVNet. 180
7.13 GUI-Marker representing the current state of a UAV. 181
7.14 TCP throughput between two nodes depending on signal strength. . . 181
7.15 UDP throughput between two nodes depending on signal strength. . . 182
7.16 RTT between two stationary nodes depending on signal strength. . . 183
7.17 TCP and UDP throughput over multiple hops. 184
7.18 TCP and UDP throughput between two distant nodes. 185

List of Tables

2.1 Comparison of evaluation in network simulation, network emulation and real world testbeds. 30

3.1 Qualitative analysis of the ADAM management architecture. 67

4.1 Possible solutions for the implementation of the virtual wireless interface. 79

4.2 VirtualMesh wireless configuration settings consisting of static parameters directly set in the simulation model *WlanModel* and dynamic parameters propagated from the virtual interfaces. 96

4.3 Qualitative analysis of simulation, testbeds, and VirtualMesh. 107

5.1 Links using 1000mW EIRP. 123

6.1 Available OViS client applications. 150

7.1 UAVNet deployment scenarios and options. 171

Preface

The following PhD thesis (accepted by the Faculty of Science, University of Bern, on 27.5.2011) is based on work performed during my employment as a research and lecture assistant at the Institute of Computer Science and Applied Mathematics (IAM) of the University of Bern, Switzerland. The research conducted in this thesis has been partially supported by the Swiss National Foundation project "Mobile IP Telephony (MIPTel)" (grant number: 200020-113677/1), the "EuQoS" Integrated Project of the European Union 6th Framework Programme (grant number IST FP6 IP 004503) and the technology transfer project "Wireless Mesh Networks for Interconnection of Remote Sites to Fixed Broadband Networks (CTI-Mesh)" funded by the Swiss Commission for Technology and Innovation (CTI) (grant number: 9795.1 PFES-ES).

I would like to thank everybody who provided me with support, ideas, understanding, and encouragement during the course of my PhD thesis. First, I want to express my gratitude to Prof. Dr. Torsten Braun, head of the Computer Network and Distributed Systems group (CNDS), who supervised and encouraged my work. He offered me an interesting and challenging work environment and the opportunity to participate in national and European research and technology transfer projects.

I would also like to thank Prof. Dr. Andreas Kassler for reading this work and providing valuable feedback and Prof. Dr. Matthias Zwicker, who was willing to be co-examiner.

Many thanks go to my colleagues at the institute and in our research group for being part of a great team. In particular, I want to thank Carlos Anastasiades, Markus Anwander, Florian Baumgartner, Thomas Bernoulli, Peppo Brambilla, Marc Brogle, Desislava Dimitrova, Kirsten Dolfus, Philipp Hurni, Dragan Milic, Benjamin Nyffenegger, Ruy de Oliveira, Matthias Scheidegger, Gerald Wagenknecht, Markus Wälchli, Attila Weyland and Markus Wulff. Special thanks go Ruth Bestgen, the secretary of the CNDS research group, for her support in the administrative tasks during all these years. She is the heart and the soul of our office.

I would also like to thank all the students, who contributed to this thesis in one way or another. In particular, thanks go to Daniel Balsiger, Reto Gantenbein, Alican

Geycasar, Adrian Hänni, Abdalla Hassan, Jana Krähenbühl, Michael Lustenberger, Simon Morgenthaler, Christine Müller, Stefan Ott, Marcel Stolz, and Roger Strähl, who performed their Bachelor's and/or Master's thesis under my guidance.

I would like to express my thanks to Paul Kim Goode, Markus Wälchli and Kirsten Dolfus for proofreading the thesis.

I am very grateful to my family, especially my parents Werner and Madeleine Staub, my brother Stefan Staub and my girlfriend Manuela Gugler, for supporting me in many ways and being always patient during the years of my PhD thesis. I would also like to thank Manuela for her understanding and sharing a wonderful time with me.

I would also like to thank my friend Philipp Berger, with whom I played squash and went to the gym for many years. This weekly activity helped in balancing the brain-centric research work.

Chapter 1
Introduction

Wireless mesh networks (WMNs) are a key technology for providing simple and inexpensive network access in scenarios where fixed (wired or cellular) network access is unavailable or expensive in installation. Such scenarios include offering public wireless network access in urban areas as well as connecting remote areas to existing network infrastructure. A key feature and requirement of WMNs is the provisioning of wireless broadband services. In addition, the application range covers wide areas requiring multi-hop routing and management.

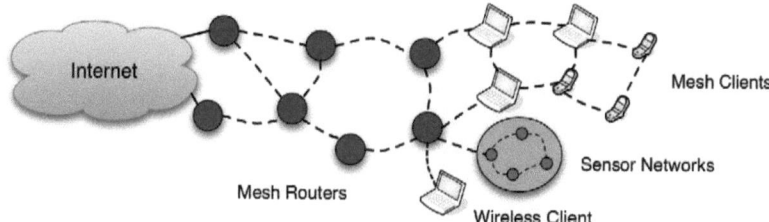

Figure 1.1: Wireless mesh network.

WMNs offer a cost-efficient last-mile access network to end user devices and sensing equipment, e.g., wireless sensor networks (see Figure 1.1). They provide means to interconnect isolated networks and to enhance wireless network coverage in urban and rural areas over a robust and redundant communication infrastructure. They usually consist of static (stationary) mesh routers and mobile or static mesh clients. Both support multi-hop communication. Depending on the scenario, clients may act as routers too.

Like any other network, a WMN and the services running on top of it traverse a given life cycle. This life cycle can be generally split into the following four phases (see Figure 1.2):

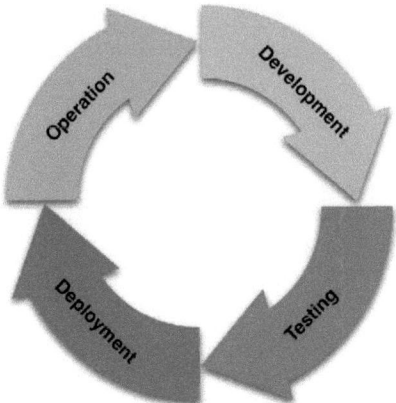

Figure 1.2: Usual life cycle of a network consisting of development, testing / implementation, deployment, and operation.

Development: In order to provide new services to the customer, network protocols and services have to be specified, implemented, and evaluated. After requirement analysis, a first implementation of the specified protocol is usually evaluated in a network simulator. Development and evaluation in a network simulator provides flexibility in terms of abstraction, cross-layer interaction, and scalability. It supports fast prototyping, enables quick introduction of new ideas, and allows the evaluation under various conditions supported by a flexible parametrisation process.

Testing/Implementation: After a successful evaluation in a simulator, the protocols and services are implemented and tested on the target platform(s). Hence, the protocol and service specifications are migrated and adapted to a real hardware platform. The migration step is in general time-consuming due to the necessity to cope with peculiarities and variations of embedded hardware platforms as well as the operating system(s) running on top of them. The migration is divided into several iterations of prototyping and evaluation. For complexity reasons, often a real small-scale testbed is used. The final release of the software is preceded by extensive pre-deployment tests in a medium-scale testbed.

Deployment: When the pre-deployment tests have succeeded, the software is released for roll-out and deployment in the target (customer) network. These steps include careful planning and scenario analysis. Environmental, social, and regulatory aspects are considered on a refined inspection. Of course, these aspects have to be taken into account from the beginning of the life cycle, since fundamental failure in satisfying them would negatively affect the success of the entire deployment. From a technical point-of-view, node locations need to be specified and the equipment, consisting of WMN nodes, enclosures, antennas, mounting solutions, and power supplies, needs to be arranged. Having prepared the hardware, the developed software is installed on the nodes. The network settings are configured and the nodes are deployed.

Operation: The network operation phase starts as soon as the network is deployed and functional. Network operation requires certain maintenance tasks. Emerging software bugs and the provision of new functionality require software updates for the operating system of the network nodes. Network management is necessary to monitor the systems and to reconfigure the network to meet changing requirements of users and network topology. Finally, changing requirements might demand new development phases.

1.1 Problem Statement

For wide deployment of WMNs, several challenges in each phase of their life cycle have to be solved. Concerning the development phase, engineers need to assess the limitations of a simplified development and modelling process in a simulator. Several limitations are faced at the testing phase as well. While the migration to the target platform introduces mainly technical challenges, the implementation and deployment in a small-scale testbed is intrinsically not able to cover all requirements of the target large-scale network. Challenges in the deployment phase include mechanical, organisational (node locations, regulations), and environmental conditions. Finally, during network operation, erroneous software or configuration updates can lead to costly on-site repairs.

1.1.1 Development Phase

The challenges in the development phase are related to abstraction level and accuracy of the simulation model. For example, simulation results are heavily influenced or even biased by the chosen wireless propagation model. The propagation model might be too simple for accurate modelling of the real environment. In addition, the impact of the operating system or the device drivers as well as restrictions of real node hardware, are not covered in simulations. Furthermore, exact timings of operating systems, real-world restrictions, or limited resources can only be roughly

1.1. PROBLEM STATEMENT

approximated in network simulations. This extends to cross-layer interactions that can be implemented rather easily into network simulation, but require more effort for being implemented on real systems. A real network stack and applications are typically not included into a simulation environment. For the given reasons, simulation results can significantly differ from results obtained by real application scenarios using real hardware. Therefore, possible limitations of the simulation environment need to be identified and assessed accordingly. Therefore, an early transition to a real prototype, which is then tested in a controlled environment (see Section 1.2.2), would be beneficial.

1.1.2 Testing/Implementation Phase

Concerning the testing phase, the migration to real hardware is difficult for a number of reasons. The system engineer has to consider the limitations of the target platform(s). Embedded devices are usually restricted in terms of RAM, storage, and CPU power. They further require an operating system and tools that are specially tailored to them and have their own limitations. Different CPU architectures require cross-compilation and make an additional step for setting up a cross-compilation tool-chain (cross-compiler, cross-linker etc.) necessary. On the occasion where cross-layer interactions are necessary, the system engineer needs to search for solutions to implement them on the real operating system with real drivers. The migration to real hardware can be simplified by a comprehensible cross-compilation build system for an embedded Linux operating system (see Section 1.2.1).

After implementation on the target platform, any new service and protocol requires evaluation in a testbed. Thereby, several limitations have to be faced. The number of nodes in a testbed is commonly limited, preventing scalability tests to some extent. Testing a prototype is time-consuming and error-prone since testbeds are rarely implemented in a completely isolated environment. Hence, interference caused by electro-magnetic radiation from external networks and devices, such as high-voltage power lines, smart-phones of employees, etc., have an impact on test results. External influences make debugging in a testbed difficult and time-consuming.

Due to varying conditions and unlike testing in network simulations, testing in real testbeds provides limited reproducibility of the results. The management of testbeds is demanding and iterative. Configuration errors during tests may cause nodes to become unavailable and lead to on-site repairs of the nodes. Another aspect is the complexity of setting up and testing mobile scenarios.

For the given reasons, it is challenging to use testbed outcomes for both the software development and the final network deployment. There are gaps between development and testing of protocols and services in a network simulator, re-implementing them in a small-scale testbed and the final deployment in the target network. In order to facilitate the implementation and evaluation of the final product, it would be beneficial to provide a system that combines the flexibility of a network simulation with real implementations, and that additionally provides useful means to assess the limitations of both abstraction levels compared to the target large-scale implementation. A possible solution is to replace the real wireless drivers with virtual device drivers, which then redirect all traffic to a simulation model for emulation of the wireless medium (see Section 1.2.2).

1.1.3 Deployment Phase

There are several challenges that a network engineer faces during the deployment phase. In addition to software-specific issues such as the stability of existing wireless network drivers, administrative issues such as political regulators, the identification of appropriate node sites and getting agreements with the land/building owners to place and access the network routers and clients can be cumbersome tasks. Furthermore, there are mechanical challenges (e.g., aligning the antennas), and the need to protect the nodes from environmental and meteorological conditions. The latter includes weather sealing of the nodes, lightning protection, and storm-proof fastening of the masts.

The deployment of a WMN is time-consuming and requires careful planning and expert knowledge limiting the applicability of WMNs to certain scenarios. Future deployments of WMNs would gain from a common knowledge-base of deployment experiences, best practises, and tested equipments (see Section 1.2.3).

In scenarios, such as setting up temporary communication infrastructures for construction sites or for disaster recovery, a temporary WMN needs to be deployed rapidly from scratch by a non-expert. In such scenarios, semi-automatic (guided) or even a completely automatic deployment might be beneficial. A semi-automatic can be supported by a self-configuring WMN and an electronic guide that instructs the user through the deployment process (see Section 1.2.4). Completely automatic deployment of a WMN can be provided by using flying robots carrying the mesh nodes (see Section 1.2.5).

1.1.4 Network Operation Phase

The major challenge during the network operation phase is to guarantee continuous remote access to the network nodes. Persistent network access for all network nodes needs to be guaranteed, even in the presence of faulty software updates or configuration errors. Network maintenance might be hindered in areas with restricted access (e.g., roof tops) or in hostile environments (e.g., disaster areas). In those environments, reliable operation of the network is of particular importance since on-site maintenance and repair is time-consuming and expensive. To some extent

1.2. RESEARCH CONTRIBUTIONS

and in some scenarios, robustness and reliability can be increased by physical redundancy of network nodes. However, this is not always feasible due to financial and/or political restrictions. Moreover, physical redundancy might be ineffective in the presence of a large-scale network failure. In addition to reliable operation and physical redundancy of nodes, network robustness further requires reliable and robust software and configuration update mechanisms. Minimally, network node failures due to update and maintenance mechanisms should be prevented. This is best achieved by avoiding software and configuration updates. Such a solution, however, is not feasible, since network operation requires software adaptation in order to support new services provided by the network nodes or to fix software bugs to guarantee security and correct operation of the network. Unfortunately, software and configuration updates always entail a certain risk of loosing remote access to the network nodes. Solutions that minimise this risk should always be favoured. Section 1.2.1 presents a solution based on self-healing mechanisms and a decentralised distribution of software and configuration updates.

1.2 Research Contributions

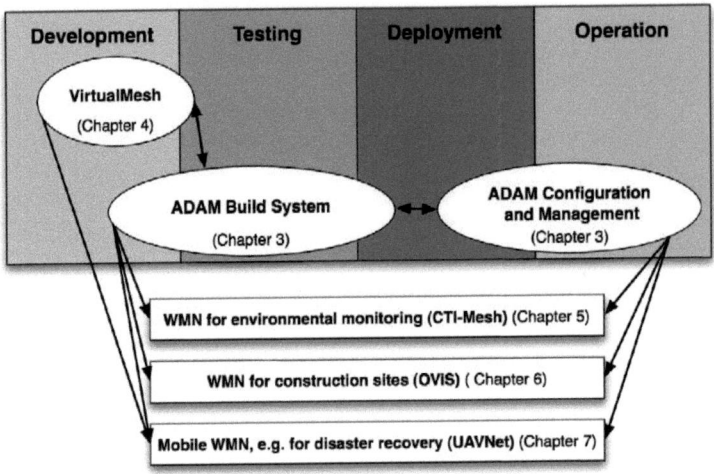

Figure 1.3: Systematic overview of contributions.

In the following, we outline our approach to provide a robust, long-term operating WMN infrastructure and software solution. A complex framework for supporting

1.2. RESEARCH CONTRIBUTIONS

a WMN during its life cycle has been developed. Figure 1.3 shows an overview of all contributions of the thesis. The components address the various challenges mentioned in the previous section, partly in functionality for specific phases and partly in designs and interactions covering multiple phases. The chronological life cycle of a WMN is shown from left to right in Figure 1.3. Solutions related to specific phases are depicted in the respective context. Testbeds and implementations, which cover aspects of the entire life cycle, are shown below the time line. The relations between the real world implementations and the supporting components (VirtualMesh/ADAM) are indicated by arrows.

The different contributions involve the following units (of functionality):

- ADAM provides a cross-compilation build system for a tailored Linux operating system, addressing the heterogeneity of WMN hardware platforms in testing/implementation. Concerning the network deployment/operation, ADAM provides management functionality to support safe software and configuration updates in a WMN.

- VirtualMesh is a module providing efficient prototyping and testing. It supports scalability and mobility tests using real implementations (based on virtualisation concepts).

- The developed components and tools have been applied to and refined in multiple real implementations, each with specific requirements. The deployment of a WMN for environmental monitoring (CTI-Mesh) both required and delivered expertise in the deployment process itself and its impacts, as well as in setting up an equipment tested in a real outdoor deployment. The OViS project required an "as easy as winking" deployment of a temporary WMN, e.g., for application on construction sites. A further specialisation and adaptation was required in UAVNet, which is based on autonomous deployment of a mobile WMN using unmanned aerial vehicles (UAV), e.g., for the purpose of disaster recovery management.

1.2.1 Operating System and Management for WMNs

Our first contribution consists of the system "Administration and Deployment of Ad-hoc Mesh networks" (ADAM). ADAM provides a build system for a customised Linux operating system, which has been ported to various embedded devices used as mesh nodes in a WMN. In addition, ADAM provides a network management architecture that safely handles software and configuration updates in a WMN - the main challenge in network operation. In contrast to existing network management solutions, ADAM provides excellent functionality to guarantee nodes' accessibility in the presence of faulty configuration and software updates. ADAM optimises the

1.2. RESEARCH CONTRIBUTIONS

availability of network nodes by using a fully decentralised software and configuration distribution approach and by several fall back mechanisms. In addition, it can cope with temporarily off-line nodes. It does require neither a co-located management network nor a working routing protocol like other management solutions. ADAM is presented in Chapter 3.

1.2.2 Development and Testing Support

The second contribution is the testing and evaluation architecture VirtualMesh, which spans the gap between development in a network simulator and testing in a real testbed. VirtualMesh addresses numerous drawbacks of both network simulations and real testbed experiments. On one hand, it includes real world parameters such as operating system timings in the evaluation. On the other hand, it provides means to diminish time-consuming prototype testing in a real testbed, the burden of setting up and maintaining large testbeds for scalability tests, and the complexity introduced by mobility tests in a real testbed. The key idea of VirtualMesh is to combine features of simulation (scalability, fast and flexible testing) and testing on real systems (operating system timings, real software). VirtualMesh implements the real network stack and application software on top of an emulated network. It emulates the wireless channel and node mobility by simulation. It can further use host virtualisation for the mesh nodes, i.e., run multiple virtualised nodes on one single host. VirtualMesh is, therefore, able to provide a virtualisation of a complete wireless mesh network. Concerning the Linux networking stack, the introduced virtual driver behaves as a normal wireless network card driver; it handles traffic redirection to the simulator instead of sending over the wireless medium. VirtualMesh is fully transparent to any software located above the virtual network driver and provides flexible testing by integrating a simulated network. Systematic and comfortable testing is provided for efficient software development as well as for extensive pre-deployment tests. This also provides benefits for subsequent real testbed implementations and the final deployment. In contrast to existing solutions, the virtual wireless device driver enabled network emulation in VirtualMesh provides a high integration of the wireless emulation, high flexibility, and good scalability at low costs. VirtualMesh is discussed in Chapter 4.

1.2.3 WMN for Environmental Monitoring

Our third contribution is demonstrating the applicability of WMNs for environmental monitoring in a rural area using the 5 GHz frequency band. The mesh routers have been equipped with directional antennas to bridge large distances of up to several kilometres. The mesh network interconnected several sensors for environmental monitoring to a fibre based backbone over a distance of more than 20 km and proved the usability and feasibility of WMNs as access networks for environmental monitoring. The deployed WMN used ADAM as operating system for the WMN nodes and benefits from ADAM's management feature to safely update configurations and software during its lifetime. The deployment activity demonstrated that such a network can be operated self-sustaining using solar-power. Our experiments provided a number of valuable contributions to the research and development community, including a documentation of extensive deployment experiences, emerged best practises, and the evaluation of appropriate tested equipment for various terrains. Our deployment activities are described in Chapter 5.

1.2.4 Deployment Support for an Ad-Hoc WMN

The On-site Video System (OViS) for construction sites is another contribution. Its key innovation is the semi-automatic deployment of temporary battery-powered WMNs by guiding non-expert users with an electronic wizard. OViS has been developed and implemented for an electric installations company. The major motivation of the electric installation company is minimising the number of on-site visits to reduce costs. With OViS, large construction sites can be supported by a video conferencing system, which reduces the need of physical presence of electrical engineers. OViS provides network connectivity even to the basements of buildings, where coverage by cellular networks is scarce or unavailable. The deployment process is semi-automatic and is assisted by comprehensible instructions to guide a non-expert user through it. The user is only required to place the network nodes according to the installation instructions displayed on a mobile client. Necessary configurations for the network are automatically handled by the system. We have developed a working prototype, consisting of battery-powered mesh nodes, which run the ADAM operating system and management, and a mobile client, which runs the deployment wizard application (electronic guide). We implemented deployment wizards for mobile clients running several major operating systems (Linux, Mac OS X, Windows) and smart-phones (Android, iPhone/iPad). Skype has been used as communication application to connect on-site users with office experts. Thus, installation issues and solutions can be discussed using an ad-hoc video conferencing system. OViS is presented in Chapter 6.

1.2.5 Autonomous Deployment of a WMN using Unmanned Aerial Vehicles

Our last contribution is a deployment framework (UAVNet) supporting an automatic deployment of a WMN using small Unmanned Aerial Vehicles (UAVs). UAVNet's main application area is in first response scenarios occurring after natural disas-

1.3. SUMMARY OF CONTRIBUTIONS

ters such as avalanches, flooding, and earthquakes. These scenarios require an automatically deployable and adaptive communication infrastructure, e.g., for video communication between the action forces. The nodes of the UAV network need autonomous location adaptation and arrangement according to the existing network requirements. Of course, these requirements can vary significantly in time given the nature of the application scenario. Most scenarios in the described applications require reliable and continuous network connectivity. In order to provide that connectivity, the UAVs have been equipped with mesh nodes communicating over IEEE 802.11s and running ADAM operating system. We have shown the feasibility of such mobile flying WMNs to connect end systems by performing a prototype implementation. The prototype WMN is remotely controlled and monitored by a user-friendly application running on iPhone/iPad devices. The development of the prototype was supported by the ADAM build system. VirtualMesh provides systematic testing functionalities for future extensions of the UAVNet prototype. UAVNet is discussed in Chapter 7.

1.3 Summary of Contributions

During the development and application of the involved projects, forming this thesis, a number of contributions have been achieved. The main contributions can be summarised as follows:

- Frameworks and tools

    - ADAM: A management framework for WMNs tailored to embedded Linux systems has been developed. ADAM includes a management architecture for fault-tolerant configuration and software updates within a WMN, guaranteeing the accessibility of the nodes and, therefore, avoiding on-site repairs. Additionally, a user-friendly and modular build system for a tailored embedded Linux distribution has been provided. It supports cross-compilation for heterogeneous WMN node platforms.

    - VirtualMesh: VirtualMesh provides a novel testing and evaluation infrastructure for WMNs and Mobile Ad-hoc Networks (MANETs). VirtualMesh exploits and applies advantages of both network simulations and real testbeds. Its key innovation is the introduction of wireless device driver enabled network emulation, where modifications of the wireless device settings are automatically propagated to the network emulation.

- WMN application scenarios with application specific tools

    - An outdoor wireless mesh network for environmental monitoring has been deployed. The deployment has shown the practicability of WMNs for large-distance environmental monitoring. It has improved the understanding of real world WMN deployments. Valuable knowledge in choosing and applying tested equipment has been gained. The gained knowledge has been made available to the research community. As a result of the deployment, useful best practises for future outdoor WMN deployments have emerged.

- OViS: A system for semi-automatic and guided deployment of a temporary WMN on construction sites has been developed and deployed. The OViS system provides a rapidly deployable communication infrastructure for online engineering support on construction sites. The OViS system has achieved a significant simplification of the deployment and application of WMNs for non-expert users.
- UAVNet: UAVNet provides the autonomous deployment of a temporary WMN using Unmanned Aerial Vehicles (UAV). It automatically establishes network connectivity between action forces in emergency and disaster recovery scenarios by a flying WMN.

1.4 Thesis Outline

The thesis is structured as follows. In Chapter 2, we investigate the work of others in relation to the developments described in the thesis. It is further helpful for understanding the main concepts of the thesis. In Part I (Chapters 3-4), we discuss our developed frameworks and tools. Chapter 3 presents our contribution concerning operating systems and management platform for WMNs (ADAM). It covers the development of an operating system tailored to WMN nodes and a fault-tolerant management architecture that improves node accessibility. In Chapter 4, we discuss VirtualMesh, a novel testing and evaluation architecture, which fills the gap between network simulation and testing in a real testbeds. In Part II (Chapters 5-7), we apply our tools to WMN application scenarios and present our experiences and developed application-specific tools. Chapter 5 describes the deployment and operation of an outdoor WMN for environmental monitoring. It documents our hard- and software setup. Gained experiences and a deployment process with best practises are examined. Chapter 6 discusses OViS, our deployment framework for temporary battery-powered WMN to perform video conferencing on construction sites. In Chapter 7, we present UAVNet, an autonomous deployment framework that provides a flying WMN for first response disaster monitoring. Finally, Chapter 8 concludes the thesis, presents further improvements, and discusses interesting and promising future directions of research.

Chapter 2

Related Work

This chapter contains background information and related work discussion, which will assist the understanding of the research presented in this thesis. First, Wireless Mesh Networks (WMNs) are introduced as one of the key technologies to provide ubiquitous network access to end users and sensing equipment. WMNs extend wireless network coverage in urban areas and offer wireless broadband connectivity to rural or developing areas, which are not covered by wired or cellular networks due to cost reasons. Other application areas are temporary networks for various purposes, e.g., disaster recovery and emergency situations.

After the general introduction of WMNs, relevant work in the area of routing, network management, operating systems, network evaluation by simulation and emulation, as well as existing WMN deployments and testbeds are discussed. Moreover, the hardware platforms that have been used for our contributions are described.

2.1 Wireless Mesh Networks

Wireless mesh networks (WMN) are evolving into an important access technology for broadband services. They provide an efficient way to interconnect isolated networks as well as to enhance the network coverage at low costs. WMNs bring us much closer to the vision of being on-line anywhere anytime.

The authors of [3, 4, 33, 160] provide an overview of WMN technology and its applications. WMNs are based on wireless ad-hoc networks. They consist of two node types: mesh routers and mesh clients. Both support multi-hop communication and may act as routers. Additionally, a mesh router may be equipped with multiple radio interfaces based on the same or different wireless access technologies. The mesh routers are rather static than mobile and form a wireless mesh backbone. They offer gateway and bridge functionalities to other networks. Mesh clients are mobile or static devices, which connect over multi-hop communication to a mesh router. They normally are more sensitive to power consumption than static mesh routers, which

2.1. WIRELESS MESH NETWORKS

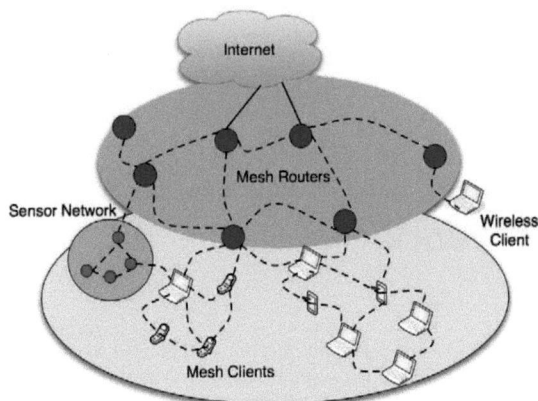

Figure 2.1: Hybrid wireless mesh network.

may be directly connected to the electricity network. The classification of WMNs leads to three main network categories: infrastructure, client, and hybrid meshes. Infrastructure WMNs build a wireless backbone for conventional clients. Community and neighbourhood networks can be built using this type of mesh networks. Client WMNs offer peer-to-peer connectivity among client devices and are similar to mobile ad-hoc networks (MANETs). The third category is formed by hybrid WMNs that are in fact a combination of the other two types and are the most commonly used WMNs (see Figure 2.1). The most important characteristics of a WMN are:

- Multi-hop wireless communication: WMNs are based on multi-hop communication to extend the network coverage and still efficiently the channel capacity at the same time. Moreover, clients without direct-line-of-sight to the network access point can be connected over multiple hops.

- Support for ad-hoc networking: In order to provide easy deployment, increased flexibility and adaptability, WMNs use ad-hoc routing mechanisms and ad-hoc connectivity. Routes within the network are automatically established and do not require manual configuration.

- Self-(*) properties, such as self-configuration, self-healing and self-organisation: Besides ad-hoc networking, WMNs include additional self-(*) properties to reduce the deployment and maintenance effort and to improve the fault-tolerance.

- Mobility depending on the mesh node type: Mesh clients can be mobile whereas mesh routers, forming the wireless backbone, are usually rather static.

- Multiple types of network interfaces: In order to provide access to the Internet and other fixed and cellular networks, mesh nodes can be equipped with various network interfaces. The mesh nodes then provide gateway functionality between the network technologies.

- Power constraints depending on the mesh node type: Mobile mesh clients can be sensitive to power consumption as they are battery-powered. In contrast, usually mesh routers do not have strict power constraints if they are either directly connected to the electricity grid or solar-powered.

- One or multiple radios per node: In order to increase the capacity, mesh nodes can be equipped with multiple radios. This enables the usage of different orthogonal channels leading to an improved usage of the available frequency spectrum and less contention on a given channel.

- Heterogeneity of radios: Mesh routers can be equipped with different types of radios. For example, the wireless back haul network can use IEEE 802.11n radios or WiMAX radios whereas IEEE 802.11a/b/g is used for the access of mesh and conventional clients.

- Compatibility and interoperability with existing wireless networks: WMNs are based on existing radio technologies and have to guarantee the compatibility with their specification in order to support mesh and conventional clients.

- Cost-efficiency: The usage of multi-hop wireless communication in WMNs avoids the expensive set up of wired or cellular infrastructures. Due to their self-(*) capabilities, WMNs can further use of inexpensive commercial off-the-self (COTS) equipment to provide a cost-efficient extension of the network coverage.

WMNs are considered to be a valuable communication technology for the following scenarios [3, 4]:

- Broadband home networking (cost efficient "last mile")
- Community and neighbourhood networking
- Enterprise networking
- Metropolitan area networks
- Transportation systems
- Building automation

2.1. WIRELESS MESH NETWORKS

- Health and medical systems
- Surveillance systems
- Emergency/disaster systems
- Vehicular networks, i.e., wireless multi-hop networks on board of trains, buses, ships, or air planes

2.1.1 Routing

The most important aspect for easy deployment and self-configuration of WMNs is ad-hoc routing. There exist several ad-hoc routing protocols, which can be categorised in single-path and multi-path routing protocols according the number of established paths between a source and a destination. In the following, routing metrics are explained first as they are necessary for route selection in any routing protocol.

Routing Metrics

If the routing decision is based on a simple hop count metric, the network performance might be reduced due to the selection of lossy links. There is a need for advanced routing metrics. Expected Transmission Count (ETX) [60] assigns each link a metric that represents the estimated number of transmissions of a packet before its successful reception. ETX is the sum of all link metrics on the route. Long routes and routes with lossy links obtain only bad grades.

Although ETX performs better than shortest path routing based on hop count, it does not take the heterogeneity of multiple radios into account. ETX only considers loss rates on the links and not their bandwidth. Therefore routes based on few long-range radio links are preferred to routes with more short-range radio links, even if they have less bandwidth and limit the spatial reuse [67]. For example, ETX favours routes with IEEE 802.11b links (max data rate = 11 Mbps) to routes with 802.11a (max data rate = 54 Mbps) links. Links with IEEE 802.11b radios are selected by the routing scheme as IEEE 802.11b radios usually have a longer range than IEEE 802.11a radios. This reduces the achievable throughput of the network.

A significant improvement of ETX that considers the bandwidth of the links is the Expected Transmission Time (ETT) metric [67]. It is based on the expected transmission time of a fixed size packet on a link. This time depends on the link bandwidth and loss rate.

Weighted Cumulative Expected Transmission Time (WCETT) [67] extends ETT to take the channel diversity into account. MR-LQSR (Multi-Radio LQSR) is an

2.1. WIRELESS MESH NETWORKS

enhancement of Link Quality Source Routing (LQSR) that uses the WCETT metric. The authors have implemented and tested the routing protocol in the Mesh Connectivity Layer (MCL). ETX and WCETT are sending probes at a fixed rate to determine the loss rate. Unfortunately, the loss rate also depends on the data rate used. The authors of [111] address this problem by comparing the measured loss rate with multiple probing rates supported by the wireless technology. The channel diversity component of WCETT does not consider "allowable" spatial reuse of channels and punishes all links sharing a channel equally, even if they are enough physically separated to avoid interference. Therefore, the performance of WCETT degenerates with increasing network size [157].

Airtime link metric c_a [95] is defined as amount of channel resources consumed by the transmission of a test frame of size B_t over a particular link (see Equation 2.1). The airtime is calculated by using the data rate r currently employed for the transmission of the test frame and the frame error rate e_f, i.e., the probability of a transmission error. The airtime link metric considers different radio technologies by specific overhead constants O.

$$c_a = \left(O + \frac{B_t}{r}\right) \frac{1}{1 - e_f} \qquad (2.1)$$

Single-Path Routing

In contrast to MANETs, WMNs provide infrastructure support, have usually less mobility and different power-constraints, introduce different a hierarchical network, and support multi-radio/multi-channel communication. Nevertheless, WMNs and MANETs share some common characteristics such as ad-hoc networking and self-organisation. Therefore, some single-path MANET routing protocols such as Ad hoc On-demand Distance Vector Routing (AODV) [133], Dynamic Source Routing (DSR) [99], Destination-Sequenced Distance Vector Routing (DSDV) [146], Topology Broadcast based on Reverse-Path Forwarding (TBRPF) [133], and Optimized Link State Routing (OLSR) [54] are often used in deployments of WMNs.

The above-mentioned routing protocols can be categorised in reactive and proactive (table driven) schemes. In reactive (on-demand) protocols, a route is only established if it is required for data transfer. This reduces the control overhead and saves bandwidth and energy during inactivity periods, but the network may suffer from significant delays until a valid route is established. If a node wants to transmit data, it has to request first a route to the destination by transmitting a route request (RREQ) message to the network. The destination or intermediate node can then reply with a route reply (RREP) message. AODV [133] and DSR [99] are typical examples for reactive routing protocols. AODV is a distance vector protocol. Therefore, a routing entry does not contain the complete route to the destination but

2.1. WIRELESS MESH NETWORKS

only the next hop, i.e., only partial routing information. A routing entry contains destination, next-hop, cost metric for the complete path, and a sequence number. If a route to a destination has not been used or re-activated for a certain period of time, a routing entry is automatically removed. In contrast, DSR includes the complete route information to reach the destination in the routing table and each packet. Intermediate nodes cache learnt routes in order to reply directly on RREQ messages.

Proactive (table driven) protocols update their routing information independent of the traffic by periodically transmitting topology control messages. Routes are, therefore, always available for data transmission. This results in lower latency than observed in reactive protocols, but produces a high overhead to keep the network topology information consistent on all nodes. Typical representatives are DSDV [146] and OLSR [54]. In DSDV, nodes periodically broadcast their complete routing tables. Routes are calculated based on the Bellman-Ford algorithm. Sequence numbers in the routing entries guarantee loop-freeness.

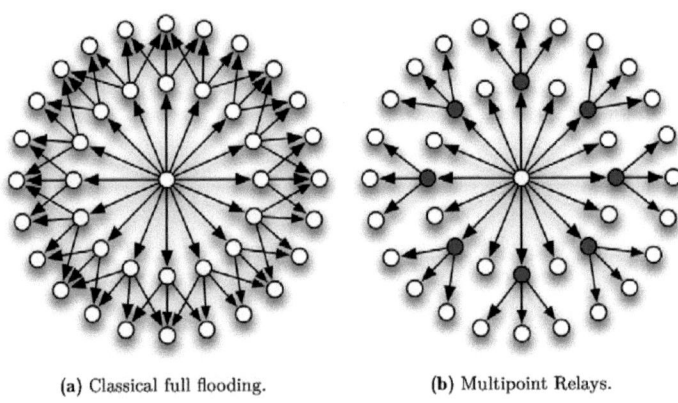

(a) Classical full flooding. (b) Multipoint Relays.

Figure 2.2: Optimised flooding OLSR: classical vs. multi-point-relay flooding.

OLSR [54] is a proactive link state routing protocol. It uses HELLO and Topology Control (TC) messages to discover and then disseminate link state information. All nodes periodically broadcast HELLO messages including their neighbour list. Thus, all nodes are aware of their two-hop neighbourhood. Each node then selects a minimal set of Multi-Point-Relays (MPR) to reach all two hop neighbours. A control message is then only rebroadcasted if it has not been received before and if the node belongs to the MPR set (see Figure 2.2). OLSR uses Host and Network

Association (HNA) messages to disseminate network route advertisements in the same way TC messages advertise host routes. Like the above-mentioned routing protocols, the original OLSR uses the hop count metric for routing. The commonly used open source implementation of the OLSR [187, 188] *olsrd* uses the ETX routing metric. OLSRv2 [53] incorporates ETX as standard metric.

The work-in-progress standard IEEE 802.11s [41, 92, 95] represents a completely different routing approach. The mesh routing is performed on the MAC layer (layer-2 routing), whereas AODV, DSR, DSDV etc. use routing on the IP layer (layer-3). Consequently, an IEEE 802.11s mesh is completely transparent for the IP layer, i.e., IPv4/IPv6, Address Resolution Protocol (ARP) and Dynamic Host Configuration Protocol (DHCP). The Hybrid Wireless Mesh Protocol (HWMP) [12] is defined as default routing protocol for IEEE 802.11s. It provides a reactive protocol for mobile networks based on AODV with a radio-link aware metric. In addition, HWMP includes pro-active tree-based routing by periodic root announcements.

In our network for environmental monitoring and the temporary WMN for construction sites, we used the *olsrd* implementation of OLSR with ETX as routing scheme as it is widely used in the community. Moreover, as our networks consist only of one type of radio and the data rates are fixed, more advanced routing metrics are not necessary. In UAVNet, we used the IEEE 802.11s mesh routing with the airtime link metric.

Multi-Path Routing

WMNs are prone to transmission failures, node failures, link failures, route breaks, as well as congested nodes or links due to the unreliability of the wireless medium, resource-constrained nodes or dynamic topology. These failures also affect the communication quality. Multi-path routing can provide a solution by establishing multiple paths between a source and a destination, providing an increased reliability and fault tolerance of the data transmission as well as load balancing.

Several multi-path routing approaches enhance the well-known single path routing protocols AODV [133] and DSR [99] with multi-path functionality. Split Multi-Path Routing (SMR) [118] is probably the most well known descendant of DSR. It extends DSR to create maximally disjoint paths. The routing scheme prohibits intermediate nodes to generate RREP messages. Intermediate nodes forward duplicate RREQs if they arrive on a different link than the one(s) already seen and if their hop count is equal or lower than the one of the already seen RREQs. The destination answers the first route request with a RREP. This represents the shortest delay path. Then the destination selects from the later arriving RREQ a maximally disjoint path to it. Both paths are used equally for data transmission.

AOMDV [122] and AODVM [211] are multi-path routing protocols that are based on AODV. AOMDV discovers multiple disjoint and loop-free routes in a single route

2.1. WIRELESS MESH NETWORKS

discovery. In order to keep track of multiple paths, the routing entries contain a list of the next-hops along with the corresponding hop counts. All next hops have the same sequence number. Furthermore, a node maintains an advertised hop count for each destination, which is the maximum hop count in all paths. Each duplicated route advertisement received by a node defines an alternate path. To guarantee loop-freedom, a node only accepts paths with hop counts lower than the advertised hop count for that destination. The neighbour list and advertised hop count are reinitialised when a route advertisement with a higher sequence number is received.

In AODVM, intermediate nodes do not drop duplicate RREQ packets; the redundant RREQ information is stored in a RREQ table at the intermediate node. The destination answers each received RREQ (from different neighbours) with a RREP. Nodes on the path overhear the RREPs. If a node is assigned to a route, it is deleted from its neighbours' RREQ tables. This method finds node-disjoint paths.

The authors of [184] propose a Quality-of-Service (QoS) extension for SMR. It satisfies multiple QoS requirements by adaptively using forward error correction (FEC) and multi-path routing mechanisms. It can fulfil a delay, a delay and bandwidth, or a packet loss requirement. This is achieved by adjusting the number of used paths, the parity length of the FEC, and the traffic distribution rate on each path. This computation considers local link information from all links, which build up the available paths. It can be executed either by the source or by the destination. If it is done by the destination, the required local link information is collected in the RREQ, else it is collected in the RREP. Each node is responsible to collect or predict (both mechanisms are possible) the local link information and add it to the RREQ and the RREP respectively.

Resilient Opportunistic Mesh Routing (ROMER) [212] is another routing solution for wireless mesh networks. It directly uses the path diversity of multiple routes to enhance the robustness of the routes. ROMER builds up a runtime forwarding mesh, centred on the long-term minimal cost path between the source and the destination on per-packet basis. The forwarding mesh offers actual candidate routes. ROMER selects the actual highest-rate link and randomly other currently available high-rate links and delivers the data packet redundantly on these paths.

In our opinion, multi-path routing could significantly enhance the reliability and robustness of WMN deployments, e.g., for environmental research. Therefore, we evaluated multi-path routing in student projects [80, 88]. Testing our Linux implementation of AODVM [139, 181] showed that AODVM, like AODV, suffers from the existence of communication gray zones [119]. In such zones no data communication is possible, although the HELLO messages indicate neighbour availability. Moreover, our tests showed that network should have a certain size in order that multi-path routing provides benefits. Otherwise, no alternative link or node-disjoint paths exist.

Therefore, we did not use multi-path routing in our small-sized networks.

2.1.2 Multi-Channel Communication

The network capacity of a WMN can be significantly increased by communicating over multiple non-interfering channels [13], resulting in an improved usage of the available frequency spectrum. Multi-channel communication enables parallel transmissions in the network and increases multi-hop communication performance by reduced interference. Moreover, the contention on a single channel is reduced.

In order to employ multi-channel communication, WMN can be equipped with multiple radio interfaces and the channels have to be properly assigned to these interfaces. Existing channel assignment strategies can be classified as *static*, *dynamic*, or *hybrid* depending on how frequently the channel assignment is performed [57, 114].

Static Channel Assignment

When using a *static* channel assignment, all network interfaces are bound to fixed channels either permanently or for long periods of time. In order to guarantee network connectivity, neighbouring nodes require at least one interface to be tuned to a common channel. If a node has fewer network interfaces than available channels, a static assignment cannot efficiently use the available frequency spectrum. The *static* channel assignment is, for example, applied in the Hyacinth testbed [151, 152].

Dynamic Channel Assignment

Dynamic channel assignment allows the nodes to frequently switch their interfaces from one channel to another. All available channels may be selected dynamically for communication, even if nodes hold less interfaces than available channels. If two nodes want to communicate, they need to ensure that they switch their interfaces to a common channel before. Two possible coordination mechanisms are *split phase* and *hopping*. Both require time synchronisation among the nodes.

In the *split phase* approach, the time is split into a control and a data phase. During the control phase, all nodes switch their interfaces to a common default channel. By exchanging control messages, the nodes then agree which channels are used during the data phase. Example protocols are Multichannel Access Protocol (MAP) [48] and Multichannel MAC Protocol (MMAC) [165].

Hopping protocols can either use a common channel hopping sequence or independent hopping sequence. In common hopping, all network nodes follow a common channel hopping sequence for the control traffic. After having exchanged control messages on the current channel of the hopping sequence, a communication pair leaves the common hopping cycle for the data transmission. Examples for

2.1. WIRELESS MESH NETWORKS

common hopping protocols are Hop-Reservation Multiple Access (HRMA) [185], Channel Hopping Multiple Access (CHMA) [192] and CHMA with Packet Train (CHAT) [193]. Independent hopping protocols divide the time in slots like common hopping approaches, but the nodes then follow independent channel hopping sequences. These sequences are learnt by the neighbours during overlaps in their hoping sequences. If a node wants to transmit data, it follows the hopping sequence of the receiver. An example is Slotted Seeded Hopping (SSCH) [11].

Hybrid Channel Assignment

A *hybrid* scheme combines *static* and *dynamic* channel assignment strategies. It requires at least two radio interfaces per node. Static (or semi-dynamic) channel assignment is used for the fixed interface, dynamic channel assignment for the switchable interface(s). Dedicated control channel protocols use this hybrid approach. They assign the fixed interfaced to a common control channel, on which then the channel assignment of the other interfaces is coordinated. Examples are Dynamic Channel Allocation (DCA) [209] and DCA with Power Control (DCA-PC) [210].

Hybrid Multichannel Protocol (HMCP) [113, 114] introduces a different hybrid channel assignment scheme. Each node statically assigns a channel to one of its interfaces (fixed interface), dynamic channel assignment is used for all other interfaces. Fixed interfaces are used for receiving data whereas the dynamic interfaces are used for sending data. Each node announces the fixed channel through periodic HELLO messages. A sender then dynamically switches one of its interfaces to the discovered fixed channel of the receiver. After having discovered all fixed channels, no channel coordination is needed anymore.

Net-X [50, 51, 115] is a framework for multi-channel communication support on Linux systems by introducing a channel abstraction layer. It uses the HMCP routing protocol. Net-X was developed for Linux 2.4 kernels. In cooperation with Karlstad University, we ported the Net-X framework to recent Linux 2.6 kernels [130]. Karlstad University further integrated OLSR with Net-X [45] and replaced the standard scheduler of Net-X by a QoS-aware one to prioritise voice over IP traffic [46]. The Net-X framework represents a possible extension for OViS (see Chapter 6).

Adjacent Channel Interference

In the described multi-channel protocols, orthogonal, i.e., non-overlapping, channels are used to avoid interference and to increase the aggregated network throughput. It is assumed that IEEE 802.11b/g offers three orthogonal channels whereas IEEE 802.11a/h offers 11, 12, or 13 orthogonal channels depending on country-specific regulations. However, measurements showed that network performance may still be degraded significantly due to adjacent channel interference (ACI) [7, 49]. ACI is caused by out-of-band radio leakage of wireless transceivers due to imperfect filters. Transmissions on neighbouring channels in the frequency spectrum may interfere with each other due to this power leakage. It is, therefore, recommended to use separated channels, i.e., leave out some channels in the assignment. According to [47], the network throughput for a given channel separation is only predictable for a data rate of 6 Mbps in IEEE 802.11a. When using higher data rates or automatic rate control, ACI and resulting network throughput cannot be reliably predicted based on the channel separation.

2.1.3 Network Management

A WMN, like other networks, requires management functionality such as monitoring, reconfiguration, and software updates. In the case of WMN testbeds, this management functionality can be established *out-of-band* by an additional infrastructure, e.g., wired or wireless back-haul networks. For a productive/operational network, this is neither feasible nor desirable. Management functionality for productive WMNs has, therefore, to communicate *in-band*, i.e., using the same communication network for management and data transmission.

Besides the Simple Network Management Protocol (SNMP) [43, 44], existing management approaches tailored for WMNs include JANUS [154], DAMON [149], MeshMan [10], MAYA [121], ATMA [150], and Abaré [147, 148]. Some provide only monitoring functionality, whereas others are full management solutions.

Monitoring

JANUS [154] is a fully distributed agent based monitoring architecture using a peer-to-peer overlay network for communication. Its architecture is similar to SNMP. JANUS has been developed for Windows-based WMNs using the Mesh Connectivity Layer (MCL) [126]. Besides missing management capabilities, the current implementation, based on a standard JAVA virtual machine, cannot be run on resource restricted devices, such as small embedded devices.

The Distributed Ad-hoc Network Monitoring (DAMON) [149] provides distributed monitoring of multi-hop networks using agents collecting relevant data. It only provides monitoring and depends on AODV routing. Therefore, it is not suitable for general-purpose WMN scenarios.

Meshman [10] is a management architecture providing an SNMP replacement that considers network dynamics in WMNs. By combining source routing with a hierarchical address scheme, it is independent of the routing scheme used. Unfortunately, the current implementation only provides information retrieval, but no configuration.

2.1. WIRELESS MESH NETWORKS

Full Management Solutions

MAYA [121] is based on OpenWrt (see Section 2.2.2) and AODV routing. It provides mechanisms to configure multiple selected nodes either over remote secure shell or by sending an encrypted UDP packet. It relies on a working routing protocol and cannot handle nodes off-line during configuration time. As a fact, MAYA cannot change the used routing protocol.

ATMA [150] is a management framework for wireless testbeds. It deploys a parallel multi-hop WMN to provide out-of-band centralised management of the actual testbed to guarantee the accessibility of the testbed device. The co-located management WMN uses Linksys WRT54G wireless routers with OpenWrt as operating system. Routing is performed by a modified version of AODV. ATMA is based on a client-server architecture. An agent is running on each ATMA node of the management network and provides auto-configuration of the node. It first assigns a temporary IPv4 address from the auto-configuration IP address range (169.254.0.0/16) [52]. Then it connects to reachable wireless networks until it receives a beacon from a central management server. If the ATMA agent has discovered a management server, it automatically performs a registration. Henceforth, the management server can configure the testbed device that is connected the ATMA node. ATMA can guarantee accessibility to the testbed devices in any circumstance. It can easily deal with misconfigured nodes in the testbed. Although, the *out-of-band* management approach of ATMA might be reasonable for testbed management, it is not suitable for productive network. Nevertheless, the boot strapping of the ATMA agent provided helpful insights for our own work.

Abaré [147, 148] provides a software-assisted process for installation and management of a WMN. It is based on a central database to co-ordinate the management. The firmware is delivered individually to each node. It cannot cope with misconfigured nodes.

There are several standardised approaches to manage and update broadband equipment of operators. Control And Provisioning of Wireless Access Points (CAPWAP) [39, 40, 134] is a standard for managing multiple wireless access points by a centralised controller. It targets large deployments of conventional wireless single-hop networks, but not multi-hop networks such as WMNs. TR-069 [26] is a SOAP based standard for management of end customer devices, e.g., DSL routers. It provides auto-configuration, service provisioning, software update, and monitoring by a central auto-configuration server. TR-069 clearly targets wired networks and is not directly applicable to wireless multi-hop networks.

Cfengine [34] is a powerful framework for network and system management. It uses distributed agents to perform policy-based network and system administration tasks. These agents periodically pull policy-based specifications of maintenance tasks from their neighbours and apply them independently. It is also possible to simulate a push method by invoking the pull mechanism remotely. This simulated push could reduce the propagation time of an update. *Cfengine* offers a lot of flexibility by its concept of dynamic grouping of nodes into classes, to which then certain policies are applied. Classification is performed fully distributed on each node. The resulting class membership defines all other actions. Although, *cfengine* has been proposed for fixed networks, the distributed agent concept fits perfectly the management requirements of WMNs. Therefore, we used *cfengine* for implementing our management architecture (see Chapter 3).

As humans are prone to errors, erroneous configurations and faulty software updates may be applied to a WMN destroying the accessibility of individual nodes and resulting in costly on-site repairs. An ideal management architecture for WMNs should, therefore, guarantee the availability of the network independently of configuration errors and faulty software updates and should use in-band communication. None of the existing approaches currently provides this functionality.

2.2 WMN Nodes

WMN nodes are usually built upon embedded hardware platforms and run an operating system, specially tailored for these resource restricted embedded hardware platforms.

2.2.1 WMN Hardware Platforms

Several commercial hardware platforms are available for WMNs, e.g., PCEngines, Linksys, Cambria, Avila, Tropos, and Cisco. WMN nodes can be deployed in various environments. The nodes should be energy efficient. Outdoor deployments may further require weather protection against weather influences, e.g., additional weatherproof enclosures. WMN nodes are usually realised as embedded systems, i.e., computer systems designed to perform few dedicated functions. Especially for community networks, the devices have to be priced low-cost. This section introduces four wireless mesh node platforms that have been used for our research work.

PC Engines WRAP

The PC Wireless Router Application Platform (WRAP) [143] is an embedded board based on an x86 compatible CPU. The boards used for our testbed are the WRAP2.C and its RoHS (EU restriction of the use of certain hazardous substances in electrical and electronic equipment) compliant successor board WRAP2.E. These boards contain a 233 MHz AMD Geode SC1100 CPU (fast 486 core), 128MB RAM, CompactFlash card slot for the secondary storage, one Ethernet port, two miniPCI sockets,

2.2. WMN NODES

and one serial port. A battery can be added to power the real-time clock (RTC). We have preferred WRAP to any Linksys Router based solution with OpenWrt due to its ability to carry two wireless miniPCI cards. This enables multi-radio/multi-channel communication. In our work, we used two IEEE 802.11a/b/g cards. The boards are sold for a competitive price (100 USD).

Figure 2.3: PCEngines WRAP.2C with an indoor case.

PC Engines ALIX

The PC Engines ALIX series [144] are a higher performance replacement for the WRAP sold for the same price. The system's design is quite similar. The ALIX.3d2 boards used in our testbed, for the outdoor deployment (see Chapter 5), and in OViS (see Chapter 6) have an x86 compatible 500 MHz AMD Geode LX800 CPU, 256 MB RAM, a CompactFlash socket to be equipped with a exchangeable storage card, one Ethernet port, two miniPCI sockets, one serial port, and two USB ports. An RTC battery can be added. The AMD Geode processor contains a hardware watchdog, i.e., a timer that reboots the node if not periodically reset. This helps in recovering a node from a non-responsive state (self-healing).

Meraki Mini

Another type of node is the Meraki Mini [125]. It is a much smaller device than the WRAP/ALIX board. It is built upon a System on a Chip (SoC) module from Atheros and contains a 180 MHz MIPS 4KEc CPU, 32 RAM, 8 MB NAND storage and a wireless interface. Most parts are integrated in the SoC. The node further provides an Ethernet port and an internal UART serial port. It has low energy consumption. Meraki is a spin-off company of the MIT Roofnet project. We have

2.2. WMN NODES

Figure 2.4: PC Engines ALIX.3d2 system board.

Figure 2.5: Meraki Mini with an indoor case.

selected the Meraki Mini as a platform for our work due to its competitive price, which is crucial for community and neighbourhood networks. The Meraki Mini has been priced at 50 USD at commercial launch, but the price has been significantly increased afterwards and we switched to the OpenMesh platform.

OpenMesh Mini and OM1P

The OpenMesh Mini [136] is a low cost wireless mesh router built upon the same hardware than the Meraki Mini, which is an Atheros AR2315(A) SoC with 180 MHz MIPS 4KEc CPU, 32 RAM, and 8 MB NAND storage. It includes an IEEE 802.11b/g wireless interface (Atheros RF2316). In contrast to a Meraki Mini, Open-Mesh OM1P further contains a hardware watchdog at the same costs. The device

2.2. WMN NODES

is, therefore, a real replacement for the Meraki Mini. It contains an internal UART serial port. The used OpenMesh OM1P nodes for UAVNet due to the device size and low energy consumption (see Chapter 7).

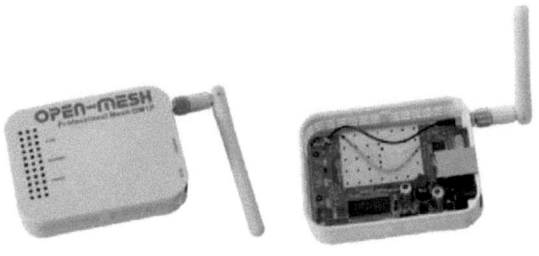

Figure 2.6: OpenMesh professional router OM1P.

2.2.2 Embedded Operating Systems Distributions

Linux, a standard operating system for personal computers and servers, is also widely used as an operating system for embedded systems; such as routers, access points, and wireless mesh nodes. There are two main Linux distributions and build systems for embedded devices, namely OpenWrt [14] and OpenEmbedded (OE) [37].

The OpenWrt [14] Linux distribution is tailored for embedded devices. Instead of the standard GNU C library (Glibc) [123], it uses the small C library replacement μClibc [5] to reduce the footprint of the compiled software, which is desirable for embedded devices. The standard UNIX tools (e.g., sh, cp, mv, grep, sed, and awk) are replaced by the multi-call binary BusyBox [6], which provides the same functionality but with a smaller memory footprint. OpenWrt uses a package manager based approach for software installation on the nodes. This provides a high flexibility for customisation of individual nodes with existing packages, but requires a read/write file system on the secondary storage. Often more RAM memory than secondary storage is available on inexpensive wireless mesh nodes (e.g., OpenMesh OM1P with 32 MB RAM and 8 MB flash storage). In this case, OpenWrt provides less software and functionality than our own Linux distribution (ADAM), which uses a compressed read-only software image. Software packages in OpenWrt can only be retrieved from a central instance, to which each node requires a connection during the update. There is no support to get cached updates from the neighbours. If software has to be updated on all nodes, each node individually fetches this update from the central server. In contrast, ADAM can retrieve software and configuration updates directly from neighbour nodes. ADAM further separates node specific configuration and node-type specific software to guarantee the same software level on all nodes and to efficiently distribute updates. This cannot be implemented using OpenWrt. Although, most users of OpenWrt download pre-compiled software packages from the software repositories, one can also build its own software package by the freely available and open source build system of OpenWrt. OpenWrt supports the cross-compilation of many software packages for a wide variety of embedded devices. The OpenWrt team provides patches for cross-compilation and the μClibc support of various software packages. We used some of these patches in ADAM.

The second Linux distribution for embedded devices, OpenEmbedded [37], is a collection of recipes for the BitBake [38] tool to automatically compile and install packages for an embedded Linux system. In OpenEmbedded, the customisation of the compilation and installation process is highly flexible. A separation of binaries and configuration data, therefore, could be implemented. Unfortunately, OE is difficult to understand due to its complexity. Another drawback is the poor support of μClibc [5] for building compact software images. Ångström [107] is a user-friendly OE distribution, but it suffers from the same drawbacks of the package manager based software installation as OpenWrt.

A different approach for building a Linux system is described in the manual Linux From Scratch (LFS) [19]. It provides systematic/step-by-step instructions to manually build a Linux system from the available sources. All instructions are well documented and alternatives are explained. Therefore, a user can customise every aspect with these instructions and explanations. Cross-compilation aspects are handled by the subproject Cross Linux From Scratch (CLFS). It provides an excellent documentation for the process of building a cross-compiler using the GNU Compiler Collection (GCC) [79]. After the installation of operating system headers, the following components are installed and compiled one by one: machine-specific Executable and Linkable Format (ELF) binary tools, intermediate cross-compiler, target C library, and final cross-compiler. Being a collection of documentation and software patches, the major disadvantage of CLFS is the missing automated build process. Nevertheless, instructions and cross-compilation patches found in CFLS helped us in the development of our own cross-compilation build system (see Section 3.3).

2.3 Network Simulation and Emulation

In the development and evaluation of new communication protocols and services for wireless networks, network simulation and network emulation are commonly used before testing in a real testbed. Whereas network simulation has its focus in

2.3. NETWORK SIMULATION AND EMULATION

	Network Simulation	Network Emulation	Real world testbed
Simplifications	- high abstraction	+ definable	+ none
Reproducibility	+ easy	+ easy	- difficult (interferences)
Fidelity	- low	+ high	+ real system
Scenario setup	+ easy	+ easy	- complex
Scalability	+ high	+ medium - bad (hardware-based)	- bad
Network traffic	- modelled	+real or modelled	+ real
Support of mobility	+ easy	+ easy	- difficult
Duration	- variable	+ soft real-time	+ real time
Costs	+ cheap	+ cheap (software-based) - expensive (hardware-based)	- expensive

Table 2.1: Comparison of evaluation in network simulation, network emulation and real world testbeds.

fast-prototyping of new ideas and their evaluation using abstract models, network emulation is used to test the real implementations in a controlled environment. Table 2.1 summarises briefly the different evaluation methods for WMNs [31, 108].

2.3.1 Network Simulation

Network simulation describes an evaluation process that first implements an abstract model of a network, then executes this model on a computer system and analyses the received output. It perfectly supports fast-prototyping of new ideas, protocols, and architectures. Network simulation provides several advantages for the design phase of new protocols and network mechanisms. First, the implementation on a network simulator is much simpler than the implementation and deployment of a real system. Second, the behaviour of the new protocols and architectures can be directly investigated under various conditions including node mobility, at a large scale and in a repeatable manner. Network simulation provides the developer an immediate feedback of his/her design decisions and parametrisation, without implementing a prototype on a real system and its time-consuming and costly testing in a testbed. Moreover, simulation supports the analysis at different abstraction

2.3. NETWORK SIMULATION AND EMULATION

levels of complex systems, to which category a computer network certainly belongs. A modular approach helps in understanding complex systems by dividing problems into smaller comprehensible tasks and modelling only the necessary parts. A model at a high abstraction level is easier to understand and to analyse than a model that already contains all details, but it can still provide reliable insights on the system's behaviour. The level of details may be increased stepwise in subsequent simulation models (top-down approach).

Although abstraction helps in the design process, it also represents a risk that the simulation model does not reflect the complete system behaviour. In general, the results of a simulation should be interpreted with care and depending on the desired goal. Simulation models are often not realistic enough to provide results fully at instruction execution level, or with high fidelity radio or power consumption characteristics. The simulation models provide their own (simplified) implementations of network protocols and applications. However, a real computer network is a highly distributed system, consisting of independent nodes. A simulation model often fails in considering all aspects of the operating systems, e.g., timing or hardware drivers.

Despite the drawbacks, network simulation is a valuable tool for fast-prototyping in the development of new protocols and architectures. Commonly used discrete-event network simulators are ns-2 [194], ns-3 [116], OMNeT++ [195, 196, 197], GloMoSim [213], and QualNet [158].

The network simulator OMNeT++ [195, 196, 197] is open-source and freely available for personal and academic use. Its implementation follows the object-oriented programming scheme using the programming language C++. It is highly modular and very well structured. Even core components, such as the event scheduler, are pluggable and can be easily exchanged. Moreover, OMNeT++ provides a convenient and rich graphical user interface (GUI), which directly visualises the simulation run. It significantly simplifies development and debugging by offering introspection of all objects at any time during the simulation run. OMNeT++ includes several simulation models for wired and wireless communication with a vast support for mobility models and communication protocols. Hence, we have selected OMNeT++ for VirtualMesh (see Chapter 4).

Direct integration of real network stacks into a network simulator is an interesting approach to increase the fidelity of the simulation and to test complex protocol behaviour [27, 98]. OppBSD [27] integrates the TCP/IP stack of FreeBSD in the network simulator OMNeT++. The Network Simulation Cradle [98] project provides support for using the real network stacks of Linux, FreeBSD, and OpenBSD with the network simulator ns-2 [194]. The integration of real TCP/IP stack provides results that are closer to a real world network. Nevertheless, OppBSD and Network Simulation Cradle do not support the testing of native unaltered applications, e.g., secure remote shell or Skype.

2.3. NETWORK SIMULATION AND EMULATION

2.3.2 Network Emulation

Whereas simulation abstractly models a complete network, network emulation [23, 24, 83, 101, 110, 155, 203, 208, 214] just duplicates the characteristics of the underlying network, e.g., by the use of simulation or system-level implementation. The network emulator then interconnects real systems by applying these characteristics to the network traffic. Network emulation is valuable for network research and protocol development, as it can approximate the real environment more accurately than pure simulation, e.g., by considering processing delays introduced by applications, operating system, and hardware. In [97] the authors validated the wireless model in the network simulator ns-2 [194] by comparing measurements of a real network setup with an emulated and simulated network. They concluded that with a proper parametrisation the simulation model can approximate the real network, but some aspects like delays introduced by hardware and operating system cannot be considered in the simulation. Their emulated network, however, provided results that matched the real measurement more accurately than the simulation.

Besides the high fidelity of the results, network emulation further offers good reproducibility and easy set up of different test scenarios including mobility. In the following, several approaches are discussed regarding their scalability, the accuracy of the wireless model and the ability to modify the wireless setup during the test.

Combining Network Emulation and Host Virtualisation

The combination of host virtualisation and network emulation also used in our own work has been proposed by [71], [110], and [215] to increase scalability. These three approaches are explained in more detail in the following.

The approach presented in [71] tries to integrate the behaviour of the real network stack and the operating system into the testing process by using virtualised hosts connected through an emulation framework. The virtual hosts are running a L4 microkernel on top of a real-time kernel. To integrate the wireless network behaviour, the hosts are connected by the 802.11b network emulator MobiEmu [214]. The wireless interface driver has been modified to communicate with the emulator instead of the physical interface, but keeping the interface to the applications unaltered. A drawback of the approach is inherited by the use of MobiEmu, the communication is either possible without errors or not at all. It does not model signal propagation and communication errors. Unfortunately, no results about the accuracy of the setup are available. It also does not propagate configuration settings of the wireless device.

UMIC-Mesh [215] is a hybrid WMN testbed. Besides a testbed with real wireless mesh nodes, UMIC-Mesh provides virtual nodes by using XEN [17] virtualisation. The virtual nodes are interconnected by a combination of advanced networking features of the Linux kernel. This includes packet filtering for controlling the com-

2.3. NETWORK SIMULATION AND EMULATION

munication between the nodes. The virtual network is only intended for software development and functionality validation. Therefore, the behaviour of the wireless medium has not been modelled in this approach.

JiST/MobNet [110] provides a comprehensible Java framework for simulation, emulation, and real world testing of a wireless ad-hoc network. It allows running the same tests independently of platform and abstraction level. MobNet is a wireless extension on top of the Java in Simulation Time (JiST) simulator [18]. The drawback of this approach is that most communication software and network protocol stacks are written in C/C++ and not in Java and, therefore, a further transition to a real-world system may be necessary afterwards.

The newly developed network simulator ns-3 [116] also adopted the concept of host virtualisation. It allows the integration of virtualised nodes running native applications and protocol stacks under the Linux operating system. The virtualised nodes in ns-3 are connected through a TUN/TAP device of the Linux kernel and a proxy node to the simulation. However, there is no support to modify device parameters of the simulation directly and dynamically by the virtualised nodes, especially for wireless devices. A similar approach for traffic redirection and internal representation of virtualised nodes has been selected and extended in our work (see VirtualMesh in Chapter 4). Our extensions include the support of direct manipulations of the wireless device parameters through usual system tools (e.g., iwconfig) and the propagation of dynamic parameters during the emulation.

Synchronised Network Emulation

When injecting real network traffic into a network simulator, there is always the problem that the simulation may not keep pace with the real network. The simulation may be too slow. In order to cope with the problem of a simulator overload during network emulation, the concept of synchronised network emulation [198, 200] has been introduced. It replaces real hosts with virtualised hosts using XEN. A central synchronising component then controls the time flow of the virtual hosts by an adapted scheduler for XEN. It keeps them synchronised with the network simulator OMNeT++ [195]. Synchronised network emulation represents a valuable extension to avoid scalability problems and could extend our own emulation approach, described in Chapter 4.

SliceTime [199, 201] is a platform for accurate network emulation of wireless networks. It picked up our approach of a virtual wireless driver (Chapter 4) and provides wireless driver-enabled synchronised network emulation for ns-3.

2.3. NETWORK SIMULATION AND EMULATION

Off-line Emulators

Off-line network emulators, such as W-NINE [145] and QOMET [23, 24], handle the real-time requirement by off-line processing of a previously generated communication scenario. They use a two-stage approach. The first stage is completely off-line. A discrete event simulator converts the wireless scenario into a time-series of network states. This state description is then delivered to the wired network emulator Dummynet [155] in order to emulate the wireless link between end points. The end points are standard computers emulating the wireless nodes. The communication is sent over Dummynet using a wired network. QOMB [22] extends QOMET to support multi-hop communications.

W-NINE, QOMET and QOMB provide repeatability and testing of real application software. They provide high accuracy of wireless properties as sophisticated models can be used in the off-line stage. Off-line network emulators cannot be used if wireless parameters are modified during the course of the emulation, e.g., by application or by operating system feedback. They are, therefore, not suitable for testing software that influences the wireless interface of a node, which would be required to support full testing of advanced schemes for WMNs. Thus, off-line emulation is not considered in our own work.

FPGA-based wireless emulators

Another approach to satisfy real-time requirements is to implement the wireless channel in hardware, by a dedicated micro-controller. The authors of [29, 103] propose such a wireless emulator using a hardware channel simulator. The unaltered network nodes are packed in radio frequency (RF) shielded boxes and their radio interfaces are connected to the hardware channel simulator, which then emulates the signal propagation using a field programmable array (FPGA). The channel simulator supports directional antennas and mobility. The system presented in [29] supports 15 nodes operating in 2.4 GHz ISM band. The main advantage of an FPGA-based wireless emulator is the provided repeatability in combination with a real MAC layer and a realistic physical layer supporting multipath fading. The main drawbacks are the high costs for RF-shielding and appropriate FPGA and the limited scalability as only few nodes can be supported by a single FPGA. Instead of the costly hardware-based emulation, software based emulation has been selected for our own work as it provides more flexibility and reduced costs due to the usage of commodity hardware.

In summary, a good approach for the development of new protocols and services is to start with testing an idea in simulation and upon success to implement a first real prototype that is then further studied in a network emulation environment. After the evaluation with network emulation, the prototype implementation can be further tested in a real testbed or deployment.

2.4 Existing WMN Deployments and Testbeds

Existing deployments are either related to research such as Transit Access Points (TAPs) [104], MIT Roofnet [1, 2, 25], Microsoft Research [66, 67], Berlin Roofnet [166], Heraklion MESH [8], WiLDNet [142], QuRiNet [205, 206, 207], UMIC-Mesh [215, 216], DES-Mesh [28, 85, 86] and KAUMesh [62, 105] or provide public network services in a metropolitan area such as the "free the net" [124] initiative in San Francisco and the freifunk.net [74] community networks.

2.4.1 Outdoor Deployments

The authors of [104] analyse the challenges of building a broadband wireless Internet access network. They propose an architecture based on fixed, wired-powered "Transit Access Points" (TAP). The TAPs form the wireless backbone of an infrastructure WMN. The "Technology for All" (TFA) network [42] represents a deployment of a WMN as a cost efficient "last mile". In cooperation with TFA, Rice University developed and deployed a WMN in one of Houston's most economically disadvantaged neighbourhoods. It aims to strengthen under-resourced communities by providing free access to information technology, educational and work-at-home tools. The network is based on twelve mesh nodes equipped with 802.11b radios and AODV is used as routing protocol. The usage of mesh technology instead of a new wire line infrastructure reduces drastically the costs of covering the area with broadband Internet access. The residents receive an entry-level service of 128 Kbps for free. Higher service levels are available at moderate rates.

MIT's Roofnet [1, 2, 25] consists of about 50 wireless mesh routers, which interconnect Ethernet networks in apartments in Cambridge, MA, USA. It is using IEEE 802.11b radio interfaces with additional omni-directional antennas. Some nodes share their digital subscriber line (DSL) and, therefore, act as gateway to the Internet. New users can easily join the network by installing the provided hardware and software kit. Roofnet requires no configuration of the network or planning. It routes packets using a new routing protocol, which is based on DSR and employs the ETX metric instead of simple hop-count. Roofnet provides an average throughput between nodes of about 627 kbps and the whole network is well served by just a few Internet gateways.

Berlin Roofnet [166] is an IEEE 802.11-based community WMNs, similar to MIT's RoofNet project in Berlin. The nodes are operated individually by students with their own equipment. The target is to build the network in a completely self-organising/self-configuring way for inexperienced users.

Heraklion MESH [8] provides a metropolitan multi-radio mesh network that covers approximately 60 km^2 with 14 nodes in Heraklion, Greece. The nodes consist of a mini-ITX board with an x86 1.3 GHz CPU and 512 MB RAM, 40 GB HDD,

2.4. EXISTING WMN DEPLOYMENTS AND TESTBEDS

up to four IEEE 802.11a/b/g Atheros-based cards. The network uses directional antennas to bridge distances of 1 to 5 km. In order to guarantee the accessibility of the nodes, a secondary wireless node, and an intelligent remote power switch is co-located to each node. The nodes are running Gentoo Linux as operating system and use OLSR as routing protocol.

WiLDNet [142] provides low-cost network connectivity in rural and remote areas (developing regions) using WiFi-based Long Distance (WiLD) links (10 - 100 km). Two production networks are maintained in India and Ghana and have shown that the commonly used Carrier Sense Multiple Access / Collision Avoidance (CSMA/CA) scheme is not suitable for long distance links. A Time Division Multiple Access (TDMA) scheme showed significant performance improvements for the long distance links.

QuRiNet [205, 206, 207] is a WMN deployed in the Quail Ridge natural reserve in California, USA. It serves for environmental monitoring. QuRiNet uses directional antennas. Its nodes are based on embedded boards similar to PCEngines WRAP equipped with two IEEE 802.11b/g wireless cards. All nodes are solar-powered. The lessons learnt are inline with the ones described in Chapter 5.

The Mesh Do-It-Yourself guide [100] tries to apply the experiences made in MIT Roofnet, Berlin Roofnet and in the freifunk.net community network to rural Africa. It provides a step-by-step guide to setup a WMN based on the freifunk.net firmware and covers some aspects of the deployment. The book "Wireless Networking in the Developing World" [73] provides a more comprehensive guide for setting up an affordable communication infrastructure for development areas using inexpensive off-the-self equipment. The description of our WMN for environmental monitoring in Chapter 5 contributes to these efforts by additional deployment experiences and tested equipment.

2.4.2 Testbeds

Microsoft Research integrates ad-hoc routing and link quality measurements in an inter-position layer framework called mesh connectivity layer (MCL) [66, 67, 126]. MCL is a Windows driver that implements a virtual network adapter. The ad-hoc network is shown as an additional network interface to the rest of the system. Routing in the MCL is done by a modified version of DSR called Link Quality Source Routing (LQSR). It is well suited for small networks with low mobility and no restrictions on power consumption.

UCSB MeshNet [20] consists of 25 nodes distributed at the University of California at Santa Barbara. The nodes are built upon two Linksys WRT54G wireless routers that are interconnected over Ethernet. One wireless router is the actual mesh node running AODV whereas the second is used for the out-of-band management of ATMA (see Section 2.1.3).

2.4. EXISTING WMN DEPLOYMENTS AND TESTBEDS

Hyacinth [151] is an IEEE 802.11-based multi-channel WMN testbed at the State University of New York. It uses nine small form-factor personal computers equipped with two 802.11a wireless cards. The nodes run Windows XP as operating system. Two nodes serve as gateways to the wired network.

UMIC-Mesh [215, 216] consists of 51 mesh nodes spread over two buildings at the Computer Science department of the RWTH Aachen University. It uses WRAP/ALIX boards with two IEEE 802.11a/b/g cards and omnidirectional antennas. The nodes are interconnected by a wired backbone to guarantee accessibility and simple management of the testbed. Experiments are centrally managed and the nodes load their operating system over the Network File System (NFS). It is a standard Ubuntu Linux distribution. Besides to the real testbed, UMIC-Mesh includes a virtualised environment by host virtualisation using XEN [17]. In UMIC-Mesh, the connectivity between the virtual nodes is configurable, but does not provide a complete emulation of the wireless medium.

The Distributed Embedded Systems Testbed (DES-Mesh) [28, 85, 86] is a research testbed at the Freie Universität Berlin and consists of 115 mesh nodes. It uses ALIX boards with two IEEE 802.11a/b/g miniPCI cards, one IEEE 802.11a/b/g USB dongle, and omnidirectional antennas. Additionally, wireless sensor nodes (MSB-A2) are co-located at each WMN node and form a parallel wireless sensor network testbed. All nodes are interconnected by a wired backbone for management. Similar to UMIC-Mesh, the nodes load their operating system over the wired network and mount the root file system from a central server.

KAUMesh [62, 105] is a WMN testbed consisting of 20 mesh nodes deployed on Karlstad University Campus. The node hardware compromises a Cambria GW2358-4 embedded board with 667 MHz Intel IXP435 XScale CPU (ARM processor architecture), 128 MB RAM, two Ethernet interface, a CompactFlash card socket and four miniPCI sockets. The node is equipped with three IEEE 802.11a/b/g cards. The nodes are connected to a wired backbone and can be reset remotely. Network monitoring is performed out-of-band using the open source software Nagios [76]. The testbed management of KAUMesh provides node reservation and access control. It is used for research on multi-channel communication, packet aggregation and opportunistic mesh connectivity [45, 46, 47, 61, 63, 64, 65, 117, 168].

Another approach for testing real implementations in a very flexible network is provided by the ORBIT testbed [153]. It provides a configurable indoor radio grid for controlled experimentation and an outdoor wireless network for testing under real-world conditions. The indoor radio grid offers a controlled environment as an isolated network, in which background interferences can be injected. Although the 20 x 20 grid of nodes offers a large variety of different topologies, it can be too restricted and mobility tests are even more limited. Furthermore, the scarce ORBIT resources may be not available for all experiments.

2.5. DEPLOYMENT SUPPORT FOR WIRELESS MESH NETWORKS

The network testbed Emulab [204] provides various experimentation facilities with advanced experiment management controls. For experiments with wired networks, network nodes run standard operating systems (FreeBSD, Linux, and Windows XP) and communicate over an emulated network using Virtual LANs and the emulator Dummynet [155]. Emulab has been extended to the wireless domain [203] by an IEEE 802.11a/b/g testbed. Several nodes with real wireless interfaces are deployed on the floors of an office building and can be integrated in an Emulab experiment scenario. Besides the lack of mobility support, the Emulab wireless testbed suffers from limited repeatability due to the shared location in an office building with interferences from productive networks.

Limited mobility is supported in a further testbed, named mobile Emulab [101]. However, the current mobile Emulab is not suitable for IEEE 802.11-based networks. Small robots, whose movements can be remotely controlled through an Emulab control script, carry wireless sensor motes with 900 MHz radios. The current setup uses an IEEE 802.11b network for the remote control. This and the size of the testbed room limit possible extensions of mobile Emulab for WLAN experiments. Moreover, the robots-based testbed is a costly and scarce resource similar to the ORBIT testbed.

2.5 Deployment Support for Wireless Mesh Networks

A rapidly deployable WMN for first response scenarios has been proposed in [167]. It uses OLSR routing and battery-powered mesh nodes. The network is deployed by a user walking within the area that has to be covered with wireless connectivity. Therefore, the user carries an active node that constantly monitors its links to previously deployed nodes and provides user feedback. The node periodically broadcasts probe requests, which are answered by probe replies of the previously deployed nodes. Using these replies, the active node receives bi-directional signal-to-noise ratio (SNR) measurements and can monitor its network connectivity. If connectivity falls below a certain threshold, the node indicates that it has to be deployed. The focus of the work is to cover a potentially large area with wireless coverage, whereas certain scenarios, such as the construction site in Chapter 6, require that a potentially large distance is covered by the network. The proposed deployment mechanism is considered to be a possible solution in scenarios where a large area has to be covered with wireless connectivity when using robots or unmanned aerial vehicles for the network deployment (see Chapter 7).

There are several research projects that employ flying robots, i.e., unmanned aerial vehicles (UAVs), for establishing rapidly deployable communication infrastructures. Examples are AUGNet [32], SMAVNET [89, 90], AVIGLE [156], Airshield [59], and AWARE [135].

2.6. UNMANNED AERIAL VEHICLE HARDWARE

Ad Hoc UAV-Ground Network (AUGNet) [32] is a MANET consisting of nodes mounted at fixed sites, on ground vehicles, and in fixed-wing UAVs. It uses an embedded computer similar to the PCEngines WRAP with IEEE 802.11b wireless interfaces. A more advanced system is SMAVNet [89, 90], which uses a swarm of fixed-wing UAVs that autonomously establish an emergency network for disaster recovery between multiple ground users. The UAV swarm automatically adapts to the current communication needs by communication based swarming. In contrast to SMAVNet, AVIGLE [156] and Airshield [59] employ small quadrocopter UAVs to provide communication infrastructures. AWARE [135] provides middle-ware and functionalities required for the cooperation among UAVs and targets a self-deploying network by means of UAVs based on autonomous helicopters. Chapter 7 describes our own prototype of a flying WMN using small quadrocopter UAVs.

2.6 Unmanned Aerial Vehicle Hardware

Figure 2.7 shows a small unmanned aerial vehicle (UAV) that we have used for UAVNet, described in Chapter 7. The UAV is a quadrocopter from the Mikrokopter community project [35], which was started in 2006.

Figure 2.7: Quadrocopter from HiSystems Ltd. - a small unmanned aerial vehicle.

Its principle of flight is as follows: Four rotors with fixed propellers are mounted in the same horizontal plane and turn pairwise in two opposite directions, i.e., the

2.7. REGULATIONS

front/back rotors turn clockwise whereas the left/right rotors turn counter-clockwise. The rotating propellers, therefore, just produce a lift of the quadrocopter. The torsional moments annihilate and the quadrocopter goes up, hovers, or declines depending on the speed the rotors. In order to fly in one direction, the quadrocopter increases the speed of the rotor that is opposite to the wished flight direction and brings the quadrocopter in an inclined position. The quadrocopter flies in the wished direction. The quadrocopter, therefore, can fly forwards and backwards (roll), to the left and to the right (roll). In order to turn the quadrocopter around its vertical axe (yaw), a speed difference between back/front and left/right propellers is introduced, the torsional moments do not annihilate anymore resulting in a rotation of the quadrocopter.

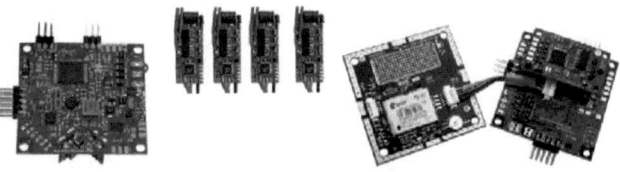

Figure 2.8: Quadrocopter flight electronics: main processor board Flight-Ctrl, four brushless controllers, GPS module, NaviCtrl with three-axis magnetic field sensor.

A Mikrokopter quadrocopter contains various control electronics to provide a stable flight (see Figure 2.8). First, the so-called Flight-Control controls the brushless engines by four brush-less controllers. It determines the current position in air by three rotation speed sensors (gyroscopes) and a vertical acceleration sensor, which measures the angle towards the earth. An optional height sensor provides relative height values. According to these sensor values, the Flight-Control adjusts the speeds of the engines and provides a stabilised flight of the quadrocopter. The NaviCtrl, a second controller boards, offers the possibility of autonomous flights by adding a GPS receiver and a three-axis magnetic field sensor (compass).

We have selected the Mikrokopter quadrocopter electronics for UAVNet (see Chapter 7) due to its good availability, moderate costs, open source software, and community support.

2.7 Regulations

Communications are usually regulated. Knowing these regulations is, therefore, an important aspect of each network deployment. In the following, we show regulations relevant for our outdoor deployment described in Chapter 5.

Alongside with other specifications, these regulations for communications limit the maximum transmit power by the equivalent isotropically radiated power (EIRP) [96]. EIRP is defined as the emitted transmission power of a theoretical isotropic antenna to produce the same peak power density as in the direction of the maximum antenna gain. It is calculated by subtracting cable losses and adding the antenna gain to the output power.

Swiss regulations released by Federal Office of Communication (OFCOM) restrict outdoor communications following the IEEE 802.11h standard to the higher 5 GHz frequency band (5.470 – 5.725 GHz) [132]. IEEE 802.11h extends IEEE 802.11a with transmit power control (TPC) and dynamic frequency selection (DFS) to cope with regulations in Europe. The effective regulations relevant for our work are listed in the technical interface specification RIR1010-04 [132], which is based on EN 301 893 [72]. They include the following restrictions:

- A maximum value of 1000 mW (30 dBi) equivalent isotropically radiated power (EIRP) is permitted with TPC. A maximum value of 500mW EIRP is permitted without TPC. With TPC, an 802.11h device shall automatically reduce its transmit power to the lowest level that guarantees a stable and reliable connection considering the expected attenuation and the variability of signal quality at the receiver. TPC results in reduced interference to other systems sharing the same frequencies. The lowest value in the TPC range of a device has to be at least 8 dB below the maximal EIRP limit.

- Dynamic frequency selection (DFS) is mandatory. It shall detect interference from radar systems, automatically switch to another channel, and, therefore, avoid concurrent operation with these systems on the same frequency. In addition, uniform spreading of the used spectrum is required.

2.8 Conclusions

In this chapter a general overview of Wireless Mesh Networks, their applications scenarios and basic concepts have been given. We introduced relevant related work in the area of routing, network management, operating systems, and evaluation methods for developments in the area of WMNs, i.e., network simulation and network emulation. Moreover, we described existing WMN deployments and testbeds, regulations and deployment frameworks.

In the next part of the thesis, we describe our general frameworks and tools, namely ADAM and VirtualMesh. The next chapter presents ADAM, which provides a management framework for WMNs and build system for an operating system tailored for WMN nodes. Chapter 4 then discusses VirtualMesh, a novel testing and evaluation architecture.

2.8. CONCLUSIONS

In Part II (Chapters 5-7), we then apply the general frameworks and tools to WMN scenarios, including a WMN for environmental monitoring, an ad-hoc WMN for video conferencing on construction sites, and a flying WMN for disaster recovery management. We further present our experiences and application specific tools.

Part I

General Frameworks and Tools

Chapter 3

Operating System and Management for WMNs

In this chapter, we describe a management architecture for WMNs and MANETs, namely *Administration and Deployment of Adhoc Mesh networks (ADAM)* [15, 16, 128, 172, 174, 180]. ADAM avoids costly on-site repairs and reconfigurations of nodes in WMNs due to misconfiguration, corrupt software updates, or unavailability of nodes during updates. It improves the accessibility of individual nodes in all these situations. Its main concept is based on decentralised distribution mechanisms for safe configuration and software updates as well as on self-healing mechanisms.

ADAM introduces epidemic distribution of configuration and software updates, i.e., nodes periodically fetch (pull) newly available updates from their one-hop neighbours. This mechanism can cope with nodes being unavailable during the update process. It provides full flexibility with a modular approach including full support of IPv6 and configuration of network services. Fall back mechanisms guarantee a node's accessibility and allow it to be recovered even from faulty software updates.

Besides the main contribution, which is a flexible and extensible framework to set up and maintain a heterogeneous WMN by safe reconfigurations and software updates, the ADAM framework provides a simple, intuitive build system for an embedded Linux distribution. This build system automates all steps required to compile and generate a Linux distribution optimised for WMNs, supporting all ADAM management features, from software source archives.

The structure of this chapter is as follows: Section 3.1 discusses the motivation for the development of ADAM. In Section 3.2, the main concept and the architecture of ADAM are explained. Section 3.3 presents the ADAM build system for an embedded Linux distribution, tailored for WMN nodes. The management operation of ADAM is then discussed in Section 3.4. After the evaluation in Section 3.5, Section 3.6 concludes Chapter 3.

3.1. INTRODUCTION

3.1 Introduction

Most existing deployments of WMNs (see Section 2.4) cover large geographical areas and include node locations that are difficult to reach, e.g., roof tops. In addition, they may be deployed in hostile environments such as desert, mountain, or arctic regions. Physical access to certain node sites may be very restricted or even impossible due to administrative or technical reasons. In general, on-site repairs are time-consuming and costly. Therefore, their number should be minimised.

During network lifetime, reconfiguration and software updates are necessary in any WMN. For example, if a security bug is discovered, a bug fix has to be installed as a software update to avoid security attacks. An example for a software update and reconfiguration is the deployment of new developments in MAC and routing protocols within the network. Unfortunately, there is always the risk that faulty reconfigurations and software updates may disrupt the nodes' network connectivity, which often results in manual on-site repairs, i.e., repairs that a network operator goes on-site to have physical access to the broken nodes.

The three main reasons for on-site repairs are modified network parameters, corrupt software updates and nodes that become unavailable during the processing of updates. First, modifications of the network parameters, especially the radio parameters such as reducing transmission power or changing the wireless channel, may drastically impact the network topology or even cause disconnection of some nodes from the network. Second, a corrupt software update may prevent a node from working correctly. Third, some nodes may be temporarily unavailable during the reconfiguration or software update distribution. Afterwards, they may not able to integrate themselves into the network due to modifications missed during their disconnection from the network. Examples are solar-powered nodes with drained batteries or transmission difficulties due to special weather conditions. Without any self-healing mechanism providing an automatic recovery, costly physical on-site access is required to repair these disconnected nodes.

The most important design consideration for a management architecture is avoiding situations such as mentioned above. Additionally, certain peculiarities of WMNs have to be considered, including limited capabilities and resources of the mesh nodes and limited network capacity for management and software updates. A mesh node is usually an embedded device without a display. It only has limited computational capabilities and a limited amount of random access memory and flash-based secondary storage. Common operating systems for mesh nodes are specially tailored embedded Linux distributions. A management architecture should be lightweight and not significantly increase the operating system's consumption of resources such a CPU time, main memory and secondary storage. Moreover, management operations, including software updates, should only use few network resources.

In the following, we use the exemplary scenario of a WMN with regular mesh

3.1. INTRODUCTION

nodes and management nodes, as illustrated in Figure 3.1. Although the WMN can be heterogeneous and consist of different node types, i.e., different hardware platforms with different capabilities, nodes of the same type usually run the same software and only differ in their configuration. As described above, some of the nodes may be temporarily unavailable. This limiting factor has to be considered in the management architecture. Management functions can be performed either by distinct management nodes or by ordinary mesh nodes. Management nodes are usually equipped with better hardware than the regular mesh nodes and can provide more features. Their primary task is the monitoring of the network as well as the configuration of all network parameters. For the ease of use, their functionalities can be accessed via a web interface. They could further include advanced tools, e.g., software image generators or a complete development environment. During the lifetime of the network, new nodes may be added to the network, others may be removed.

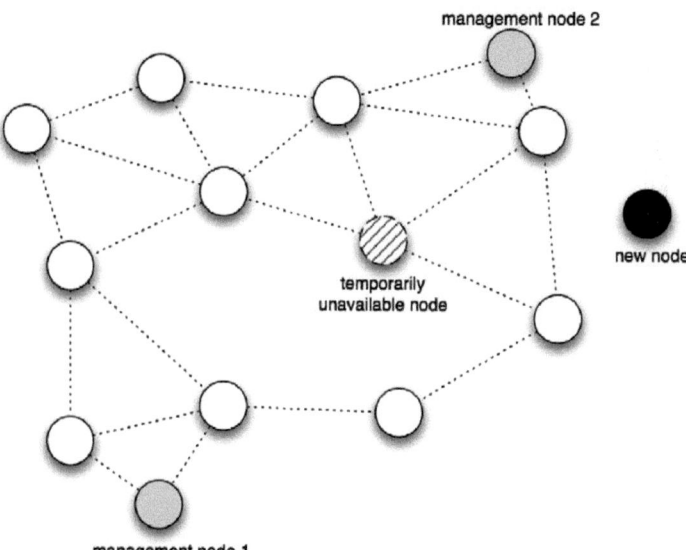

Figure 3.1: Example of a WMN: One node is temporarily unavailable, e.g., due to lack of power. Another node is added to the network for the first time. Multiple nodes can provide management functionality for the network.

A management architecture for a WMN has to fulfil the following requirements:

3.2. ADAM: CONCEPT AND ARCHITECTURE

- The management operation should be fully distributed and decentralised to prevent a single point of failures.

- Network connectivity of the nodes has to be guaranteed even in presence configuration errors and software updates in order to avoid costly on-site repairs.

- Management communication should be always encrypted and only take place between authenticated nodes to prevent malicious actions or hijacking of the network.

- As routing protocols should be configurable, the management architecture should be independent of any specific routing mechanism.

- In the on-going transition to IPv6, a management architecture for WMNs should already fully support of IPv4, IPv6 and IPv4/v6 dual stack operation to support all possible deployment options.

- To support future modifications and improvements, the management architecture should be modular and extensible in terms of manageable configuration parameters

- The management framework should be portable to support various node hardware platforms found in WMN deployments. This includes resource-constraint low-cost devices commonly used in community networks.

- The network management should be user-friendly, e.g., provide a web-based front-end.

3.2 ADAM: Concept and Architecture

ADAM meets the requirements of the described scenario by building on the three main concepts, which includes *decentralised distribution of software and configurations*, *self-healing mechanisms*, and *separation of node specific configuration and node type specific binary software images*.

3.2.1 Decentralised Distribution Mechanism

The first main concept of ADAM is a *decentralised mechanism for distributing software and configuration updates* (see Figure 3.2). Each node periodically pulls new software or configuration updates from its one-hop neighbours. Therefore, updates are epidemically propagated from one node to the other throughout the entire network. A periodicity of two minutes provides a good trade-off between management

3.2. ADAM: CONCEPT AND ARCHITECTURE

overhead and a timely propagation of the updates. The update mechanism works independent of the routing protocol used. If a node is not reachable during the reconfiguration, it fetches the updates when it is up again. If the gathered updates target the node, they are automatically applied. The successful application, as well as network connectivity, is guaranteed by self-healing mechanisms, which are the second main concept of ADAM.

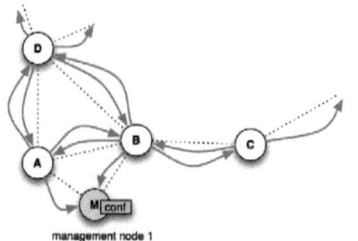

(a) Nodes periodically check for updates (green arrows). A new configuration is injected at a management node (M) or a normal node.

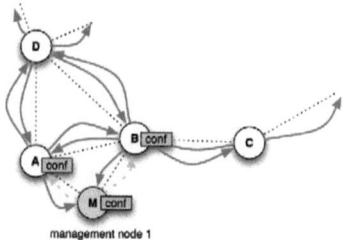

(b) First nodes (A, B) get the update from node M (orange arrows).

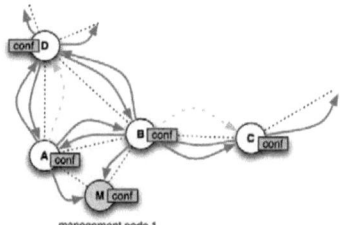

(c) Next nodes (C, D) get the update from node A and B.

Figure 3.2: Distribution of node configuration and software updates.

3.2.2 Self-Healing

The *self-healing capabilities* of ADAM are manifold. They include monitoring of the network topology during updates, detection of isolated nodes, and automatic rollback to the latest running software if a software update fails to boot properly. Monitoring of the network topology and appropriate reaction is the first self-healing

3.2. ADAM: CONCEPT AND ARCHITECTURE

mechanism. If network parameters are modified that may disrupt network connectivity, self-healing mechanisms recover the network connectivity. One example is the discovery of a reduced number of neighbours after lowering the transmission power. ADAM step-wisely increases the transmission power to the previously set value in order to reach at least predefined network connectivity. The detection of isolated nodes is the second self-healing mechanism. It supports that temporarily unavailable nodes can be reintegrated into the network, even if the network configuration has completely changed during their absence. Isolated nodes may discover their state and follow an automatic lost node procedure for re-joining the network. The final self-healing mechanism takes care of faulty software updates. Although ADAM uses checksums to detect data corruption and to guarantee error-free transmission of updates, the configured software updates may still contain errors that prevent the nodes from properly booting after the update. If such errors occur, an automatic rollback process is started. The node then automatically reboots and loads the latest known working software.

3.2.3 Separation of Software and Configuration Data

ADAM separates *software and configuration data* on a node to exploit similarities between the nodes and reduce the amount of transferred data in a network. It is not efficient to just distribute a software image for each individual node. Most software such as the operating system kernel and binaries for tools and applications are the same for similar types of nodes. Therefore, each node in an ADAM network contains two image files. One image file holds the operating system kernel and the binaries. This image is the same for all nodes of a similar type. The other image just holds all the node specific parts. These are mainly configuration files, which can vary for individual nodes. ADAM even splits up this configuration image into the normal configuration files and a special network configuration file. This network configuration file holds all dynamic network parameters, from which the normal configuration files are automatically generated. Dynamic parameters include IP settings, default routes, external DNS and NTP servers, IP forwarding and firewall rules, ad-hoc routing protocol, and the settings for services such as IPv6 router advertisement daemon, NTP and DNS running on the node. The full configuration image of an individual node with a size of 1 MB is usually not distributed. Therefore, ADAM only needs to distribute this network configuration file with a size of 10 KB for each node and the software image for each node type (<6 MB). This drastically reduces the total amount of transferred data for an update.

3.3 ADAM: Build System

No existing build system for an embedded Linux distribution (e.g., OpenWrt, OE or CFLS described in Section 2.2.2) properly supports all requirements for ADAM, e.g., splitting binaries and configuration. As none was suitable for the implementation of the ADAM management approach, we decided to develop an own build system based on the documentation of CFLS and some patches from OpenWrt.

The ADAM build system is especially tailored for WMNs and supports several target platforms. To prove heterogeneity support, we currently use nodes from three vendors, namely PCEngines ALIX and WRAP embedded boards, Open-Mesh Mini and OM1P, and Meraki Mini (see Section 2.2.1). The nodes differ in their processor architecture (x86 compatible, MIPS), their amount of RAM (32 - 256 MB) and secondary storage (8 MB - 4 GB). Despite these significant differences, all nodes provide similar functionality of installed utilities and software to the user.

ADAM provides a build system that produces software and configuration images for different node types. The operating system is a fully customised embedded Linux. It offers all key functionalities for a WMN node within a small memory footprint (< 6 MB), which is a key factor for the deployment on small embedded systems. In order to achieve this small footprint, ADAM uses the same tools as OpenWrt. The μClibc [5], that requires only 400 KB of storage, replaces the standard C library, and BusyBox [6] replaces the standard UNIX tools (e.g., sh, cp, mv, grep, sed, and awk), saving more than 4 MB of storage compared to the standard tools. In contrast to OpenWrt, the Linux kernel and the binaries are stored in a read-only compressed image on secondary storage, which is decompressed to the RAM during run-time. This results in up to 6 MB additional software packed on the OM1P (8 MB secondary storage) compared to running OpenWrt with a file system on the secondary storage.

The goal of the ADAM build system is to simplify all necessary steps for image creation. It avoids a steep learning curve for new users by an easily understandable modular approach and focusing on functionalities used in WMNs. It provides a simple and intuitive command-line interface. It is easily extendable to support additional software packages as well as to support other hardware platforms by integrating new build profiles, containing all necessary build parameters for hardware platform, such as board name, processor architecture and default software packages. Moreover, the software requirements of ADAM are moderate. A standard desktop machine with a current Linux distribution (Fedora, Ubuntu, Debian, Gentoo) providing the ordinary development and build tools, such as the GNU compiler collection (gcc) and its standard development tools, is sufficient to use ADAM.

The ADAM build system consists of two tools, namely the *build-tool* to compile the software and the *image-tool* to pack the software correctly into the images. Figure 3.3 illustrates the necessary steps to build a Linux distribution for an ADAM

3.3. ADAM: BUILD SYSTEM

mesh node. It uses colour codes for the used tools (orange, green, yellow, red). The first three steps use the *build-tool* (orange), steps 4 and 7 the *image-tool* (green). Steps 4 and 5 are either performed by the web management front-end or manually (yellow). The final installation and deployment (red) in step 8 is completely node-specific and, therefore, has to be performed manually.

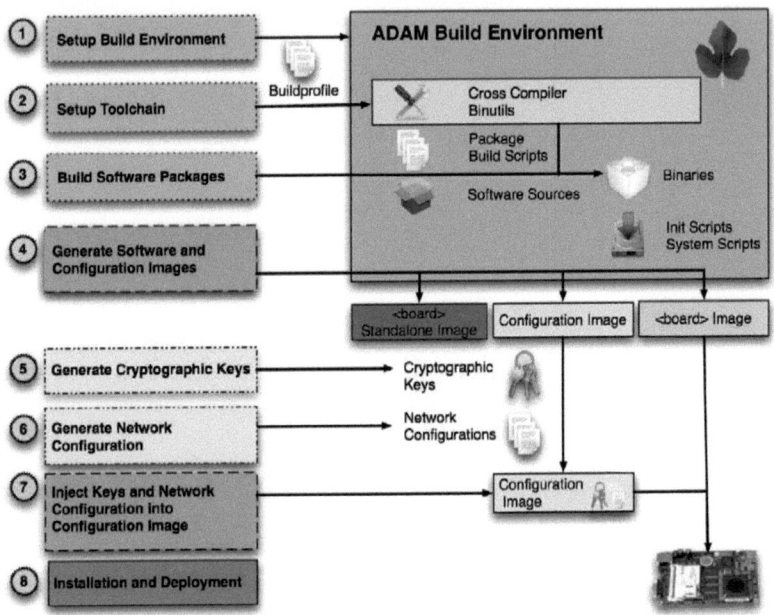

Figure 3.3: ADAM: Steps of the build and set-up process for a node.

In step 1, after installation of the ADAM build system, the set-up procedure is started by the *build-tool*. It creates a build environment for the target platform by adding a user on the local machine. The command shell environment of the new user, e.g., *alix-builder*, is set up with all necessary parameters for the cross-compilation process, such as library and compiler paths. The parameters are defined in the build profile of the selected target platform.

In step 2, the tool-chain for the cross-compilation is set up and installed for the user. The operating system headers are installed. Then, machine-specific Executable and Linkable Format (ELF) binary tools, the intermediate cross-compiler, the C

3.3. ADAM: BUILD SYSTEM

library μClibc for the target platform and the final cross-compiler are compiled and installed one after the other. The final cross-compiler is used to compile all software packages for the target platform in step 3.

An individual software package in ADAM is defined as a recipe for compilation and installation. It is implemented as simple shell script. Executing this script downloads the particular package source archive, decompresses it, applies necessary patches, configures it for cross-compiling and installs the binaries and configuration files to the correct directories after successful compilation.

In step 4, the *image-tool* is used to generate the software image for the target platform and individual configuration images for each node.

In steps 5 and 6, cryptographic key pairs for the distribution engine and the network configuration for each node are generated. The node-specific keys and the network configurations are then injected into the configuration image of the corresponding node in step 7. In the final step, the generated Linux system images are loaded onto the secondary storage of the new nodes or distributed using the ADAM distribution engine.

Besides software and configuration image, the *image-tool* creates another image type - the so-called stand-alone image. This specific image is fully self-contained and does not require any configuration image. In ADAM, it is only used for testing purposes and the installation of normal software images on the secondary storage of Meraki and OpenMesh nodes. The boot loader of these nodes does not support writing files larger than 5 MB to the secondary storage. Therefore, a stand-alone image is booted over the network to write the software images under the temporary Linux system.

Figure 3.4 shows the run-time memory layout of ADAM nodes. Depending on the platform, a node can store multiple software and configuration images. During run-time, a software image and a configuration image are mapped to a root file system. As the state, such as random seeds and log files, should not be lost, when software or configuration images are exchanged, it is stored in a special permanent storage on the node, which is also mapped to the root file system in the RAM.

Figure 3.5 illustrates the boot process of an ADAM node. After being switched on, the boot loader reads the boot configuration and loads the Linux kernel and the initial RAM based file system from the software image. During OS initialisation, the kernel then loads the root file system. After initialisation the content of configuration image is mapped on top of the root file system, the permanent storage with the node's state and log files is mounted in the system. The Linux system applies the network configuration, starts the configured system services including the time-based job scheduler. The system is now fully functional and the job scheduler periodically starts the ADAM distribution engine.

ADAM nodes load their entire file system into RAM in order to increase the sys-

3.3. ADAM: BUILD SYSTEM

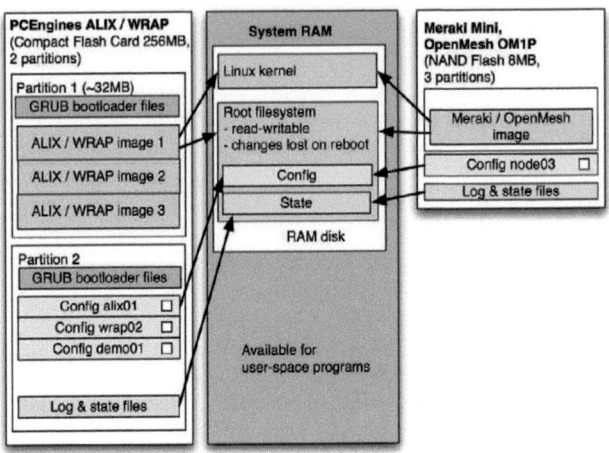

Figure 3.4: Run time layout of system RAM and the secondary storage for PCEngines ALIX/WRAP, Meraki Mini and OpenMesh OM1P nodes.

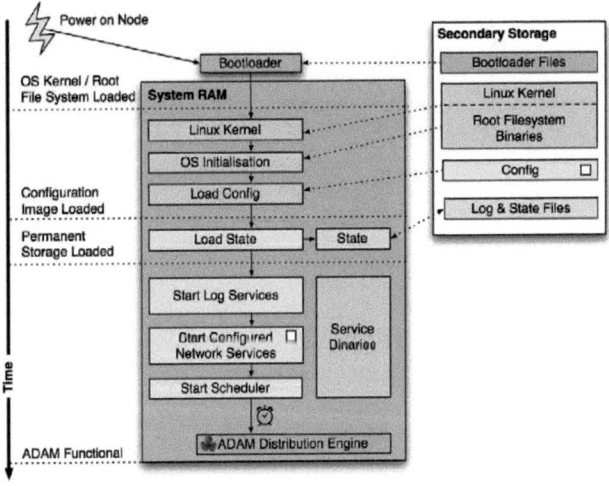

Figure 3.5: Detailed boot process.

tem performance and to take care of the limited write cycles of the secondary storage (CompactFlash cards, NAND storage). ADAM uses the *Initramfs* file system, which provides flexible memory management, i.e., its size in RAM can grow and shrink as needed. Using *Initramfs*, the entire file system is writable and individual files can be modified. These modifications are, however, not saved over a reboot due to the reload of the original software and configuration images at the next boot. Therefore, ADAM includes a procedure to write back files to the configuration image and to save them permanently over reboots.

3.4 ADAM: Management Operation

After initial installation, each node holds a software image and a configuration image with the initial network configuration. The node is physically deployed at the final location (e.g., on a rooftop). Henceforth, physical access to the node may be costly or difficult. Therefore, the node should be completely managed from remote by the ADAM configuration framework.

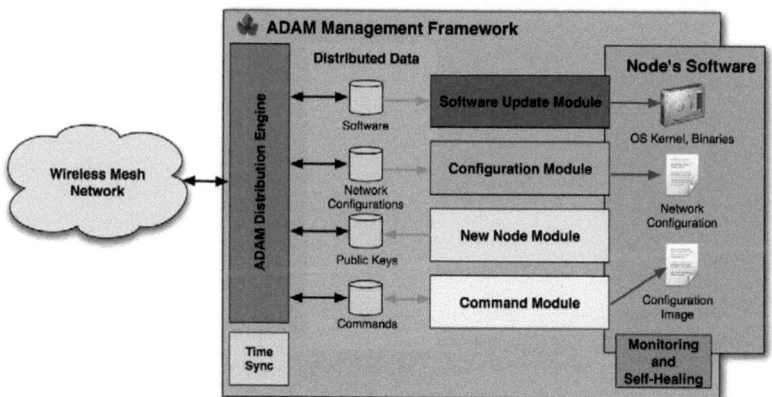

Figure 3.6: General ADAM management architecture.

Figure 3.6 shows the general ADAM management architecture. It consists of the ADAM distribution engine and modules for network configuration, integration of new nodes, software update and a generic command module. The modules are described in the following subsections.

3.4. ADAM: MANAGEMENT OPERATION

3.4.1 ADAM Distribution Engine

The ADAM distribution engine for configuration and software updates is based on the distributed management agent architecture of *cfengine* (see Section 2.1.3). It is implemented as custom scripts and policies for the distributed agents of *cfengine*. The pull based decentralised distribution can cope with nodes that are unreachable during configuration. It further guarantees encrypted and authenticated management communication, but requires time synchronisation. In contrast to *cfengine*, ADAM does not rely on an external NTP server or a battery-driven real time clock for time synchronisation. Before each distribution round, an ADAM node synchronises the system time with its neighbours by connecting to a time service running on their web servers. Each node periodically starts reachable neighbour detection, synchronises its clock, connects to all its detected one-hop neighbours and checks the availability of newer network configurations or software images. If there are updates available, they are pulled by the node. This epidemic distribution mechanism even works without a configured routing protocol.

Nodes that have been offline during the distribution of the updates get the configurations and software updates from their neighbours as soon as they are online again. If critical parameters, such as the wireless communication channel or frequency band, have been modified, a node has no connection to any of its former neighbours and can, therefore, not fetch the update. It is the automatically re-integrated into the network using the lost node detection (see Section 3.4.6).

In order to be independent of other network configuration settings, the ADAM distribution communicates over a dedicated IPv6 network using Unique Local IPv6 Unicast Addresses (RFC 4193 [93]). The dedicated IPv6 network is always present, cannot be switched off, and is only used for the management traffic. Although running on the same physical network interfaces, there is a clear distinction between this management network and the configurable IPv4/IPv6 networks used for data transmission.

Upon reception of new network configurations or new software, the node automatically applies them using the configuration and the software update module, if it is the target, i.e., the network file contains its host name, and the software image matches the node type. Otherwise, the files just remain in the exchange storage in order to be distributed to other nodes.

The ADAM security concept is based on the *cfengine* tool and employs state-of-the art authentication and encryption methods. The nodes authenticate each other based on public key – host name pairs, which are exchanged and manually approved prior to deployment. The communication is encrypted using a random session key. After initial public key exchange, the ADAM distribution engine operates under the assumption of mutual trust among all configured nodes.

3.4. ADAM: MANAGEMENT OPERATION

3.4.2 Configuration Module

The configuration module is started by the ADAM distribution engine if a new network configuration (<*hostname*>.*conf*) has been received. The <*hostname*>.*conf* file contains key value pairs for most configuration parameters, e.g., the IP address of the Ethernet interface as *eth0_IP="130.92.66.40"*. These dynamic parameters are used to generate and update most of the other configuration files. If the file name matches the host name of the node, the configuration module automatically applies the new network configuration to the node, restarts the network interfaces, and reloads all affected system services. The configuration module provides high modularity and is easily extensible. All parameters related to network configuration including common services can be set in a single configuration file. In addition, the configuration module fully supports IPv6.

3.4.3 New Node Module

The ADAM distribution engine only accepts communications from known nodes. In order to guarantee encrypted communication, it has to know the public keys of all its communication peers in advance. The new node module handles the integration of new nodes, of which public keys and network configurations are unknown within the network.

A newly set-up node includes already all configurations and keys of the other network nodes. In contrast, the already deployed nodes are not aware of the new node. The network administrator provides the public key and network configuration of the new node to the new node module, which then distributes them using the ADAM distribution engine. If the ADAM web configuration tool is used to generate the new configuration, keys and network configurations are automatically handed over to the new node module.

In ADAM, new nodes can be easily integrated into the WMN. In order to provide full flexibility and to consider all possible deployment situations, a network configuration of the new node can be directly generated and loaded on the node during its setup, preloaded and distributed in the network, or not generated until the new node tries to join the network. The new node is directly integrated into the network if it has all necessary keys and the network configuration has not changed since the node setup. If the new node holds a deprecated or no network configuration, the new node procedure is started.

Figure 3.7 depicts this new node procedure necessary for adding a new non-configured node to the network. A standard image has been loaded to the new node. Furthermore, the node has received a unique host name, its public/private key pair, as well as the public keys of the other network nodes. The keys are essential to guarantee that only authorised nodes can connect to the network.

3.4. ADAM: MANAGEMENT OPERATION

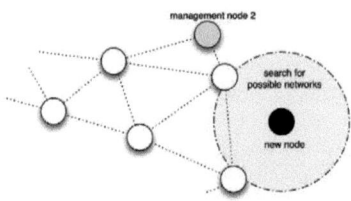

(a) New node searches for networks having an ESSID that matches an IPv6 prefix.

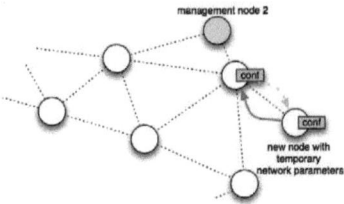

(b) New node automatically configures a valid IPv6 address and tries to get its configuration from neighbours. After the new node has received its configuration, it is fully integrated into the network.

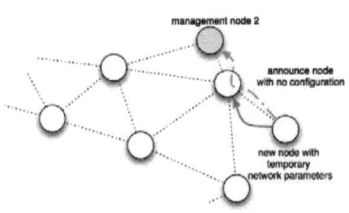

(c) If no configuration is available, the node announces its state to a management node. The user has to generate a new configuration. The new node is integrated in the network after having received this configuration.

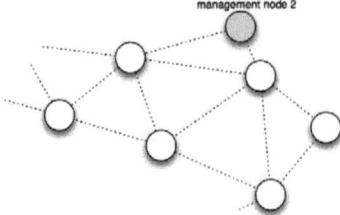

(d) After the new node has received its configuration, it is fully integrated into the network.

Figure 3.7: Integration of a new node into an existing network.

Due to the separate permanent IPv6 management network, the process of new or lost nodes joining the network is as follows. After connecting to the wireless network with the ESSID being the IPv6 network prefix, a new node contacts its neighbours for network configuration and software updates using the distribution engine as it always has an automatically assigned valid IPv6 address. If a network configuration or software update is available, the newly deployed node simply loads them from one of its neighbours, applies them, and is then fully integrated in the network. If no network configuration is available at the neighbours, the node signals its lack of a valid network configuration over the distribution engine throughout the network towards the management nodes. The user is then prompted to generate a configuration on the management node.

3.4. ADAM: MANAGEMENT OPERATION

3.4.4 Software Update Module

The software update is responsible for applying a new software image to the node. Software images are distributed together with an update file that contains detailed information about the specific update action. It contains the file name of the software image, the node type, the update version, and a checksum of the software image. If the node retrieves a new software image by the ADAM distribution engine, the software update module checks if node type, version, and checksum match. If positive, the module calls the update procedure specific for the node type. For nodes with sufficient secondary storage capacity, the safe update procedure is started. It supports the availability of the nodes even after faulty updates. Platforms without sufficient secondary storage do not support this safe update procedure. Consequently, a failed software update requires physical access physical access in case of such platforms, e.g., Meraki Mini and the OpenMesh OM1P.

The safe software update procedure of ADAM can recover from a failed software update by rebooting with the previously working software image. In order to support this functionality, a node requires sufficient storage capacity to hold at least two software images in its secondary storage. The safe software update process is based upon an additional update partition on the secondary storage and the boot loader *grub*'s ability to perform the following actions at boot time according to its configuration files:

- Install the Master Boot Record (MBR) pointer to another boot partition. The node then boots from this partition the next time.

- Boot the operating system from the current boot partition.

Employing these actions enables booting an updated kernel and performing a fall back to the previous software update when the kernel from the software update fails to boot, e.g., due to a kernel panic or a corrupt root file system.

For implementing the safe software update, a node contains two partitions. The second (update) partition holds a secondary set of boot loader configuration files that boots the current software image known to be working. In case of an update, the software update module copies the update image to the first partition and adjusts the boot entry on the first partition to boot this image at the next reboot. The node is rebooted. During start-up, the boot loader rewrites the MBR to point to the safe entries on the second partition. If the update image can be successfully booted, the update is made permanent by replacing the standard image and readjusting the boot loader files. Otherwise, a boot flag of Linux enforces a reboot, if a kernel panic occurs or the root file system could be reloaded. In this case the node automatically reboots and loads the standard image using the same boot loader configuration

3.4. ADAM: MANAGEMENT OPERATION

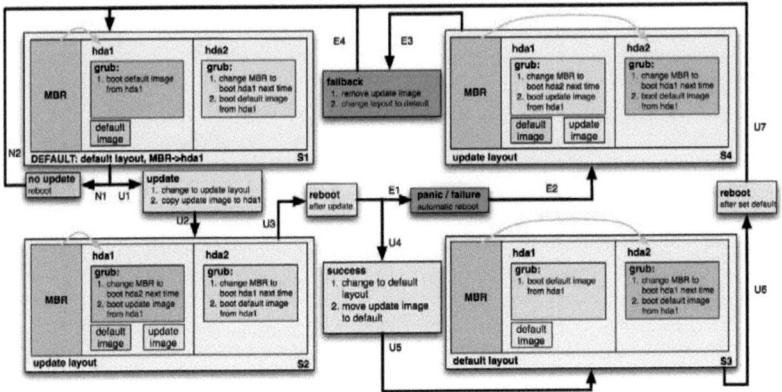

Figure 3.8: Safe software update process for Linux kernel and root file system with automatic fall back to previous software image.

on the second partition. In this way, a safe update of the software image can be guaranteed in any circumstances.

Figure 3.8 shows the detailed safe software update process. In the following, the sequences of the software process for the following three situations are:

Normal operation: The system is in default configuration (S1). No update is planned. Therefore, the system remains in default configuration after a reboot (N1/N2).

Successful update: The system is in default configuration (S1). The MBR points to /dev/hda1. The default image would be loaded after reboot. An update is intended (U1). The layout of *grub* is changed to update layout (U2). The update image is copied to /dev/hda1 (S2). As MBR points to /dev/hda1, from where *grub* configuration is read at the next reboot (U3), the boot loader *grub* sets MBR pointer to /dev/hda2 and loads the update image. If the update has been successful, the layout is reverted to the default layout (U4/U5) and the default image is replaced by the update image (S3). During the next reboot the boot loader (*grub*) configuration on /dev/hda2 is read. MBR is changed to point to /dev/hda1 again. The default image is loaded from /dev/hda1 (U6/U7). The node returns to normal operation (S1).

Faulty update: The system is in default configuration. MBR points to /dev/hda1 (S1). The update image is copied to /dev/hda1 and the update layout is

3.4. ADAM: MANAGEMENT OPERATION

set (U1/U2/S2). The system is rebooted (U3). MBR is reset to point to /dev/hda2. The update image is loaded. The update image produces a kernel panic (E1). The node is automatically rebooted (E2) and is now in error state (S4). As the MBR points to /dev/hda2, the MBR is reset to boot /dev/hda1 next time and the default image is loaded (E3). The node runs with the old kernel again. The update image is removed; the layout is reset to default (E4). The node returns to normal operation (E4/S1).

The safe update of each node concerns only the software image, which contains the kernel and the basic software of the system. Before starting a safe update procedure, ADAM checks the correctness of the boot loader configuration files and corrects them if necessary.

3.4.5 Command Module

The node-specific configuration images are not directly distributed as an entire image over the ADAM distribution engine due to the huge transmission overhead. There are usually only small changes in the configuration image. Adding, removing and modifying configuration files in the configuration image is performed by the generic command module. This module can execute user-defined commands on a predefined set of nodes. The commands are written in a command file, which is then propagated together with data files within the network using the ADAM distribution engine. Upon reception of a command file for the node, the command module executes the command and registers the execution time, the exit status, and a possible output of the command in a reply file, which is then distributed within the network. For example, if a bug fix for the file *hotplug2.rules* should be applied to some nodes, a command file with instructions for the file replacement is copied together with the new file to the exchange directory of a node. ADAM distributes the files within the network and the bug fix is applied on all specified nodes. In the example, only 10 KB of data have to be transmitted instead of an entire new configuration image with a size of 1 MB.

3.4.6 Lost Node Detection

There are scenarios, in which a node can totally loose its connectivity to all other network peers due to misconfiguration that is not properly handled by sanity checks during the updates. After a predefined time-out without any network connection, the node, therefore, resets its transmission power to the maximum value and then searches on all wireless channels for an ad-hoc network with the service set identifier that matches an IPv6 prefix. If a network is found, the node connects to it and then fetches a new network configuration (new node procedure). We recommend

3.4. ADAM: MANAGEMENT OPERATION

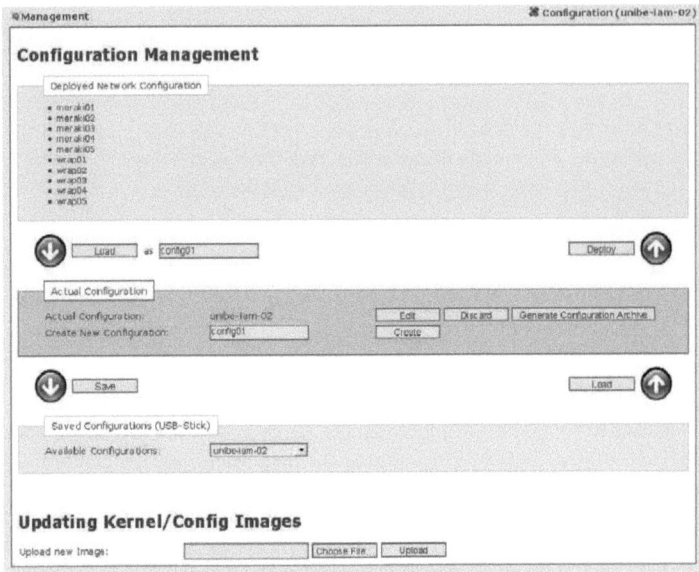

Figure 3.9: ADAM: Management of network configuration.

to configure a time-out of 2 h due to epidemic distribution mechanism and the software update time of about 50 min for the low cost nodes (Meraki Mini and OpenMesh OM1P). The lost node detection is part of the monitoring and self-healing mechanisms shown in Figure 3.6.

3.4.7 Web-Based Management

ADAM introduces management nodes to provide a user-friendly web-based management front-end that helps the user performing common management tasks, e.g., generating network configurations and network keys or uploading new images. As additional software is required for this functionality, only more powerful nodes, e.g., ALIX nodes, provide web-based management functionality. The build target *alix-mgmt* of the ADAM build system includes all necessary additions such as a web server and some scripts to the normal *alix* image.

The ADAM management front-end is illustrated in Figures 3.9 - 3.11. Figure 3.9 shows the front-end for the configuration management for an entire network. On the left top corner, the currently deployed network is shown. Then either this deployed

3.4. ADAM: MANAGEMENT OPERATION

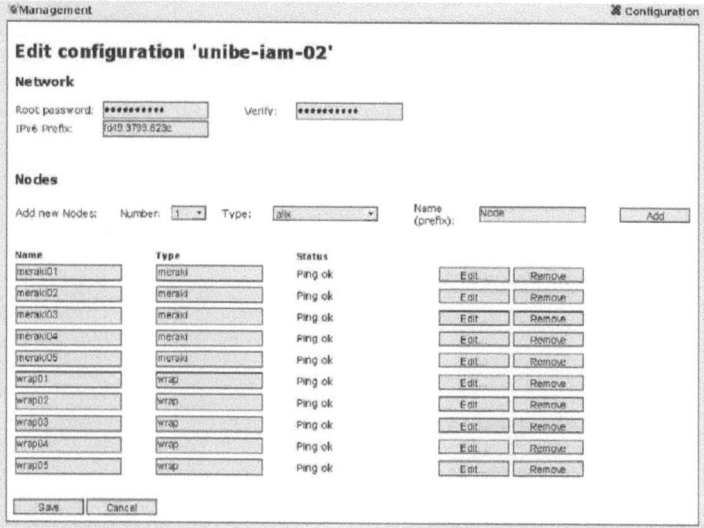

Figure 3.10: ADAM: Modification of selected network configuration.

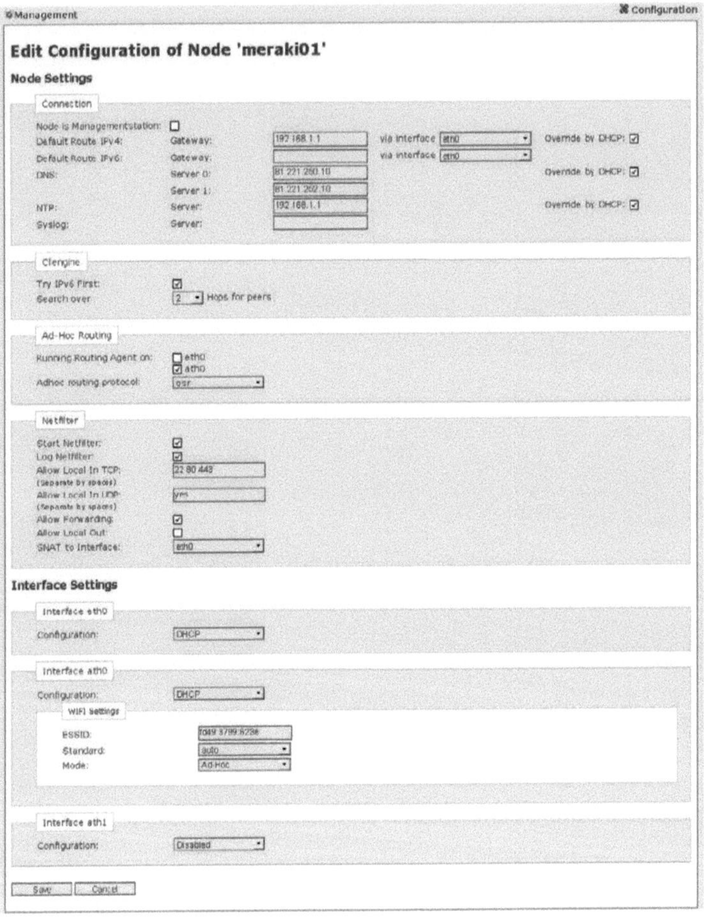

3.4. ADAM: MANAGEMENT OPERATION

Figure 3.11: ADAM: Edit the network configuration of an individual node.

configuration is loaded as actual configuration to be edited or a new configuration has to be created. After modification, the configuration is either saved to a USB memory stick or it is deployed to the network. The front-end further offers the possibility to load a configuration, which is stored on a USB memory stick. The loaded configuration can then be either modified again or directly deployed in the network.

Figure 3.10 shows the front-end for editing the configuration for an entire network. The only static parameters of an ADAM based network are the root password of the nodes and the IPv6 prefix, which is used for permanent IPv6 management network and its ESSID. The front-end shows all deployed nodes and offers their management including adding new ones. The configuration can be edited for each individual node. The various network parameters that can be modified are shown in Figure 3.11.

If a new node is added to the currently deployed network configuration, its public encryption key and network configuration are automatically transferred to the ADAM distribution engine during the deployment. All keys and network configurations can be downloaded as compressed archive from the configuration management (see Figure 3.9). This archive can then be injected into the configuration image of the new node (see step 7 in Figure 3.3).

In addition, the configuration management web front-end in Figure 3.9 offers uploading software and configuration images for deployment within the network.

3.5 Evaluation

The most crucial requirement for a management and software update architecture is to guarantee the nodes' accessibility in any circumstances, i.e., in presence of configuration errors, faulty software updates, and nodes unavailable during reconfiguration. ADAM succeeds in guaranteeing the nodes' accessibility by a decentralised software and configuration distribution and several self-healing mechanisms. A detailed qualitative analysis of ADAM can be found in Table 3.1.

The proper functionality of ADAM has been verified within our own testbed at the Institute of Computer Science and Applied Mathematics with ALIX, WRAP, Meraki Mini and OpenMesh nodes (see Section 2.2.1). Several software and configuration updates performed in the testbed proved the full functionality of both management architectures. To evaluate the times used for ADAM management, we set up a small testbed consisting of four nodes being all within a common communication range; an ALIX node, two Meraki Mini nodes, and an OpenMesh OM1P node. Using software built with ADAM, booting an ALIX node takes up to 30 s, a Meraki Mini or an OpenMesh OM1P node up to 5 min. The one-hop distribution took up to 2.5 min for a software image of 6 MB and up to 2 min for a network

3.5. EVALUATION

configuration file. Upon reception of a software update at an ALIX node, it takes about 1 min until it is reachable again, if the update could be successfully applied. If a faulty software update is distributed and then applied to an ALIX node, the safe-update procedure automatically reverts the node to the former state and the node is also reachable again after about 1 min. Due to slow write speed of the secondary storage, the update takes up to 50 min on a Meraki Mini node or an OpenMesh OM1P. In case of an updated network configuration with modified network address, an ALIX node is reachable after 30 s, a Meraki Mini or OpenMesh OM1P after 1.5 min under the new IP address. If a node is misconfigured and does not have access to the network anymore, it is automatically reintegrated into the network after 2 h as defined in the default configuration.

In order to test the command module, we set up a chain topology with the same four nodes (ALIX, Meraki01, Meraki02, OMP1). A command file for Meraki01, Meraki02, and the OM1P was injected at ALIX node. The commands were executed after 40 s on Meraki01 (1-hop), 3 min 36 s on Meraki02 (2-hop), and 4 min 54 s on the OM1P. The ALIX node received the reply files after 1 min 6 s (Meraki01), 4 min 32 s (Meraki02), and 6 min 46 s (OM1P). These times highly depend on the distribution engine, which uses a random back off time (0 - 60 s) to prevent synchronous connections of all nodes.

	ADAM
Management	
Decentralised distribution engine	+
Executable from each node	+
Support for temporary unavailable nodes	+
Self-healing mechanisms	+
Independent of routing	+
Fixed parameters	(ESSID), host name, cryptographic keys
External time synchronisation or battery-powered RTC	optional
Splitting of images	software image (< 6 MB), configuration image (< 1 MB), network configuration (< 10 KB)
Directly distributed files	software image, node specific network configuration
Modular design	+
Management framework portable	+

Additional features	- generic command module
Software image	
Linux kernel version	2.6.26 - 2.6.38
Wireless driver	Madwifi (Linux kernel version $\leq$ 2.6.28), ath5k and ath9k (Linux kernel version > 2.6.28)
Support for IEEE 802.11s	+
IPv4/IPv6 support	+ / +
Build system	
Automated build process	+
Automated image creation	+
Modularity	+
Extensibility with new software	+
Cross-compilation	+
Platforms supported	ALIX, WRAP, OpenMesh Mini, OpenMesh OM1P, Meraki Mini, XEN
Adding new build targets	+
Support for VirtualMesh (see Chapter 4)	+
Requirements for build system	desktop computer with recent Linux distribution (Ubuntu $\geq$10.10 recommended)
Delivery of the build system	download as compressed archive (1.5 MB)

Table 3.1: Qualitative analysis of the ADAM management architecture.

3.6 Conclusions

In this chapter, we proposed a novel management architecture for WMNs, namely ADAM. During the network lifetime, several management activities, such as reconfiguration and software updates, are necessary in any WMN. Misconfiguration and corrupt software updates may disrupt the network connectivity of some nodes. This leads to costly on-site access and repairs.

In order to guarantee the accessibility of individual nodes, ADAM introduced a decentralised distribution mechanism for software and configuration updates, self-healing mechanisms, and a safe software update procedure. Furthermore, it splits the node's firmware in two parts, a node specific configuration image ($\sim$ 1 MB) and a binary software image that is the same for a specific node type (< 6 MB).

3.6. CONCLUSIONS

Instead of distributing a firmware image for each node throughout the network, ADAM only has to distribute one image per node type. In order to reduce the data to be transmitted, ADAM also extracts all dynamic and configurable network parameters from the node-specific configuration image ($\sim$ 1 MB) to a single network configuration file ($\sim$ 10 KB). All other configuration files are then automatically derived from this file. Since only the network configuration file has to be distributed, the management traffic is significantly reduced.

Due to the decentralised distribution mechanism, ADAM can cope with unavailable nodes and automatically repairs configuration and software update errors. It is completely independent of a fully operable routing protocol and works completely in-band, i.e., it does not require the presence of an additional backbone network for management. ADAM just uses a separate permanently configured IPv6 network for the management running on the same physical interfaces to allow the modification of all network parameters for the data networks. ADAM fully supports the usage and configuration of IPv4/IPv6 communication.

The ADAM build system supports cross-compilation and simplifies the preparation of a customised embedded Linux operating system for several mesh node types. It is user-friendly, easy to understand and extendable. The ADAM framework is released under GPLv2 license [173].

Our qualitative evaluation shows that ADAM fulfils all requirements of a comprehensive management architecture for WMNs.

There are several further possible extensions for the ADAM framework. Self-healing capabilities can be enhanced to react faster on disruptive configuration updates. Other extensions of ADAM are an automatic dependency calculator for the software packages or a semi-automatic conversion of OpenWRT packages to ADAM build scripts and patches. Currently, we are adding ARM Cortex-A8 based Gumstix Overo computer-on-modules as a new target to the ADAM build system.

ADAM provides excellent support for the operation of WMNs by an embedded Linux distribution and a management framework. Moreover, ADAM's build system represents a valuable tool for prototype implementations on different platforms. In the next chapter, we describe a flexible framework for extensive testing of these prototype implementations in a controlled environment instead of a real testbed.

Chapter 4

Development and Testing Support

In this chapter, we describe VirtualMesh, a flexible and comprehensive framework for development and testing of new protocols and architectures for WMNs and MANETs[176, 177, 77, 179]. Its key novelty is the concept of wireless device driver enabled network emulation. VirtualMesh simplifies the commonly used procedure of development for new protocols and architectures utilising evaluation by network simulation and testing of a prototype in a testbed.

Development and testing of new software for WMNs and MANETs is generally cumbersome and split into at least two phases, which are often not unified. First, new protocols and architectures are implemented and evaluated in a network simulation environment. Second, a prototype on real hardware is implemented and evaluated in a testbed. Unfortunately, testing by simulation often requires the developer to write software that is not directly portable to testbeds and does not include real operating systems and network stacks, whereas pure prototype testing on real hardware is extremely time-consuming and expensive. Testbeds might suffer from irrepressible external interference, which makes debugging extremely difficult. Real-world testbeds usually support only a limited number of test topologies and sites. Moreover, large-scale mobility tests are generally resource intensive and impractical.

VirtualMesh provides a testing architecture, which can be used before evaluation in a real testbed or the final deployment in the productive network. It significantly simplifies the testing process by combining the strengths of the network simulation, such as the controlled environment and scalability, and prototype testing, such as using the real network stacks and real applications. VirtualMesh offers instruments to comprehensively test the real communication software, including the network stack, inside a controlled environment and under various conditions and scenarios, including node mobility.

VirtualMesh classifies as a network emulation approach (see Section 2.3.2). It is based on emulation of the wireless medium by a network simulation and the introduction of virtual wireless interfaces, which redirect the wireless traffic of native

4.1. INTRODUCTION

or virtualised nodes over the emulated wireless medium. The properties of the virtual wireless interface can be modified in the exact same manner as the ones of a real wireless interface. The modifications are automatically propagated to the simulation model and applied during runtime - a novel feature introduced by VirtualMesh.

Employing host virtualisation drastically increases the flexibility and scalability of VirtualMesh. By adding virtualised nodes to an existing testbed or even fully virtualising an entire testbed, VirtualMesh provides significantly increased scalability with reduced hardware costs and administrative effort.

Section 4.1 illustrates key problems solved by VirtualMesh. In Section 4.2, the basic concept and the general architecture of VirtualMesh are illustrated. Section 4.3 presents the communication protocol used for the connection of the nodes with the wireless network simulation. Section 4.4 explains the concept of host virtualisation. Then, the implementation of virtual wireless interface is described in Section 4.5. Section 4.6 presents the implementation of the emulated wireless medium. In Section 4.8 the evaluation results of VirtualMesh are discussed. Finally, Section 4.9 summaries our VirtualMesh activities and presents our conclusions.

4.1 Introduction

For commercial utilisation of WMNs, new communication protocols as well as new costumer services have to be developed. The development process in WMNs is typically split into evaluations by simulations and testing a real prototype in a testbed.

First, protocols and architectures are implemented and evaluated in a network simulator. Afterwards, a prototype is implemented on the target platform such as Linux and tested inside a testbed before deployment in the real network. Simulation provides most flexibility in testing. Different and large scale experiments as well as experiments with mobility of devices and users are possible. Thus, the focus here can be set on testing and debugging the functionality of the proposed protocols. Unfortunately, simulation models cannot cover all influences of the operating system, the network stack, the hardware, and the physical environment due to complexity constraints. Therefore, the transition from simulation models to the deployable solution remains challenging.

Testing the prototype in a testbed during the implementation process is time-consuming, costly, and very limited in test scenarios. Due to economic reasons, the scale of the testbeds is limited and they are often not deployed in isolated environments, which limits the reproducibility of the results. Interferences with existing networks are possible and irrepressible, which makes debugging of new protocols very challenging. Furthermore, the number of test topologies is limited and mobility tests are impracticable. Moreover, WMNs provide an even more complex testing challenge compared to simple wireless access networks. They support mobile users and high-throughput applications. Their architecture contains self-configuring and self-healing mechanisms, which have to be included in the tests. Cross-layer protocol stack interactions have to be tested in a controlled environment without any irrepressible influences. Moreover, the tests have to cover time and delay aspects of the real network stack. Not all these tests can be fully done in simulations; it is also difficult to perform them in a testbed.

We propose to use network emulation based on simulation models (see Section 2.3.2). We use the final operating software of the nodes, to replace the wireless interfaces with virtual ones, and to emulate the physical medium for gaining more control in the development process. This substantially enhances the testing process, as the real software stack may be evaluated within a controlled environment.

Our contribution is an emulation framework for WMNs called VirtualMesh, which is based on the network simulator OMNeT++ [195, 196]. This framework offers enhanced evaluation of communication software written for real and virtualised nodes on top of an OMNeT++ simulation model. Communication software can be tested without any adaptations over an emulated network using OMNeT++. VirtualMesh uses real mesh nodes with a real network stack. It intercepts wireless traffic before transmitting it over the air and forwards it to a simulation model. This simulation model offers a vast flexibility in topologies and mobility tests. It supports changing topologies and different mobility scenarios. This makes automated testing of the real communication software with a high variety of scenarios possible. In contrast to experiments in a real testbed, there are no irrepressible influences on the experiments such as interference from neighbouring networks and power lines, steel structures of buildings, or changing weather conditions. In addition, VirtualMesh facilitates the setup of large scale scenarios by host virtualisation, which has been proposed by several works presented in Section 2.3.2. As first network emulation approach, VirtualMesh has introduced the concept of a virtual wireless driver that allows the modification of the device parameters through usual system tools (e.g., iwconfig) and the direct propagation of the dynamic device parameters during the emulation. This concept has been adopted by SliceTime for the network simulator ns-3 (see Section 2.3.2).

4.2 VirtualMesh Concept and Architecture

The main concept of VirtualMesh is to intercept and redirect real traffic generated by real nodes to a simulation model, which then handles network access and the behaviour of the physical medium. The network stack is split into two parts as shown in Figure 4.1. The application, transport, and Internet layers are handled by the real/virtualised node. At the MAC layer the traffic is captured by a virtual

4.2. VIRTUALMESH CONCEPT AND ARCHITECTURE

network interface and then redirected to the wireless simulation model, the so-called *WlanModel*. The *WlanModel* calculates the network response according to the virtual network topology, the propagation model, the background interference, and the current position of the nodes. Only the MAC layer and the physical medium are simulated. All the other layers remain unchanged and work just as in a real testbed of embedded Linux nodes. Consequently, VirtualMesh requires only minimal modifications of the network stack, i.e., the adoption of virtual interfaces, and achieves a good decoupling between real network stack and the emulated medium.

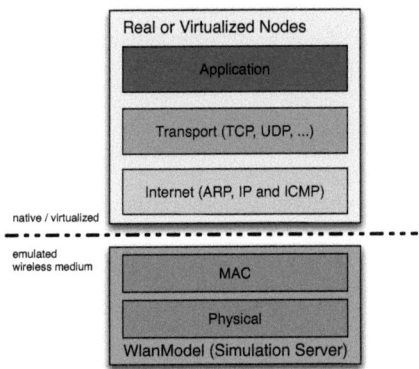

Figure 4.1: General concept: Traffic interception and emulation of the wireless medium via subdivision of the network stack.

The general architecture of VirtualMesh is shown in Figure 4.2. It consists of an arbitrary number of computers hosting the simulation model and real or virtualised mesh nodes. A wired infrastructure network interconnects the nodes and the model. The wireless interfaces of the nodes are replaced by virtual interfaces, which communicate over the infrastructure network to the *WlanModel* using the VirtualMesh communication protocol (see Section 4.3). The infrastructure network provides the communication channel between the nodes and the *WlanModel*. The entire traffic sent to the virtual interfaces is forwarded to the *WlanModel*, which processes the messages in its wireless simulation model and sends the response to the virtual interfaces of involved nodes. The network scenario in the simulation model reflects an arbitrary network topology, which is not related to the physical positions of the participating nodes. A key feature of VirtualMesh is that not only real nodes with virtual wireless interfaces are supported, but also virtualised hosts. This directly addresses the scalability problems of testbed infrastructures without additional ef-

4.2. VIRTUALMESH CONCEPT AND ARCHITECTURE

fort for the protocol developers. In our setup, host virtualisation is performed by XEN [17], but other virtualisation techniques could be used too. Host virtualisation provides additional scalability of the system. One standard server machine (Pentium D dual-core 3.2 GHz, 1 GB RAM) may hold up to ten virtual mesh nodes without any problem.

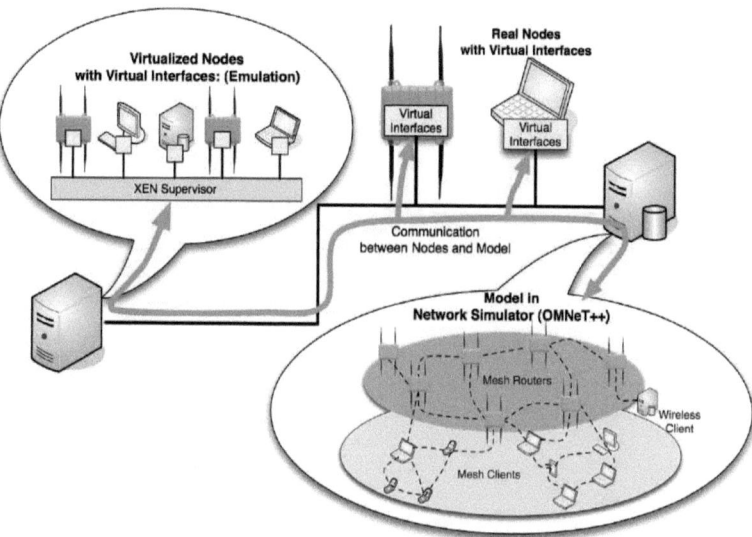

Figure 4.2: VirtualMesh architecture with real nodes, virtualised nodes, and the simulation model.

Nodes that participate as wireless nodes in the network scenario require that VirtualMesh client tools are installed. These client tools manage the virtual interfaces of the nodes and connect them to the wireless emulation (*WlanModel*). They further propagate any modification of the virtual interface settings to the simulation model. The VirtualMesh client tools are discussed in detail in Section 4.5.

The simulation server, hosting the *WlanModel*, is connected to all nodes through the infrastructure network. The server injects the forwarded traffic received from the virtual interfaces of the nodes into the *WlanModel*, which then computes the wireless propagation between the nodes and sends the correct response to the corresponding nodes. *WlanModel* is fully implemented as a simulation model for the network simulator OMNeT++ [196, 197]. It takes advantage of all features of OMNeT++ or its extension frameworks such as INET [94] or MiXiM [202]. Thus, it can make use of all different physical layer simulation models implemented in OMNeT++. In its standard configuration, VirtualMesh uses the IEEE 802.11b stack of the INET framework for modelling the wireless driver functionality. The implementation of the *WlanModel* is discussed in Section 4.6.

4.3 VirtualMesh Communication Protocol

In order to communicate between the virtual/real WMN nodes and the simulation server holding the simulation model over the infrastructure network, VirtualMesh requires a communication protocol that meets its specific requirements, such as high performance packet forwarding, dynamic adaptation of the emulated network and wireless parameter propagation. For VirtualMesh, high performance operation of the packet forwarding for exchanging the original wireless traffic is crucial. Therefore, the protocol should only introduce minimal delays in packet generation and transmission. The simulation model has to support dynamic scenarios with nodes joining and leaving the emulated network. Thus, it has to be informed about these changes timely. Any modification of a wireless interface parameter on a node has to be propagated transparently to the simulation model. In addition, the simulation model must be able to map the packets to the corresponding wireless node.

There are two possible solutions for transmitting the wireless parameters to the simulation model. All required parameters of the virtual wireless interface either are piggybacked with each forwarded wireless frame, or dedicated configuration messages are exchanged for the wireless configuration. There are several drawbacks of piggybacking. Including the wireless parameters in each data packet does not only increase the transmission overhead, it also introduces significant delays to the packet processing in the simulation model. The simulation model has to process the wireless configuration for each packet. As the wireless configuration usually remains unchanged during the transmission of multiple packets, the second solution with dedicated messages has been selected for VirtualMesh. Configuration messages are only exchanged in case of modified settings. To reduce complexity, configuration settings are sent uni-directionally from the nodes to the simulation model. Fully featured feedback of configuration parameters from the simulation model back to the nodes remains a possible extension for future work.

The VirtualMesh communication protocol uses the User Datagram Protocol (UDP) as transport protocol due to the low-latency requirement. The connectionless datagram service of UDP provides a better performance than the Transport Control Protocol (TCP). The drawbacks of UDP such as no sequence guarantee and no retransmission in case of errors can be accepted in VirtualMesh, as interconnections of the nodes with the simulation server over the dedicated infrastructure network can be considered as optimal. In any case, even some lost data packets

4.3. VIRTUALMESH COMMUNICATION PROTOCOL

would have a lower impact on VirtualMesh than many delayed packets. In summary, UDP provides the best match considering the traffic pattern of the transported wireless frames. The VirtualMesh communication protocol, therefore, transmits its five protocol messages over UDP. These message types are described in the following.

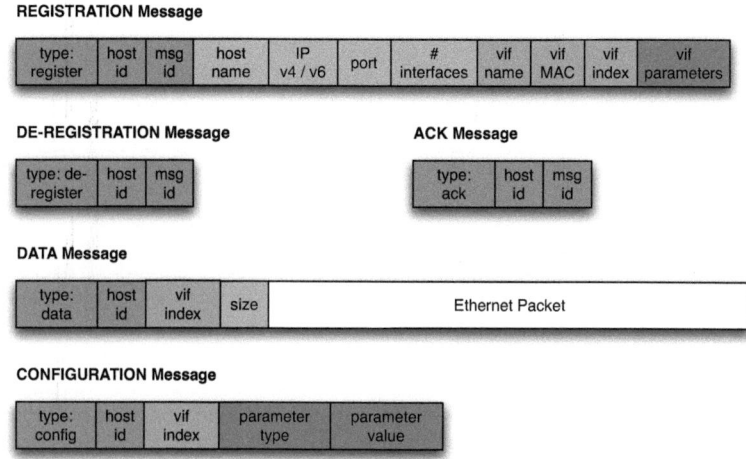

Figure 4.3: Message format to communicate with the model: data transmission and node registration.

Figure 4.3 shows the five message types of VirtualMesh necessary for node management, traffic tunnelling, and propagation of configuration settings. Messages are *REGISTRATION*, *ACK*, *DE-REGISTRATION*, *DATA*, and *CONFIGURATION*.

At start-up, each node intending to participate in the wireless emulation of VirtualMesh registers itself by sending a *REGISTRATION* message to the simulation server. This message contains the host identification (unique id, e.g., the uniquely assigned MAC address 000b6bdbe502), a sequence number (msg id), the host name (e.g., node01), the infrastructure IP address (IPv4 or IPv6), the port where the VirtualMesh client tools are listening for incoming traffic, and the number of interfaces. It further contains the initial values for all dynamic parameters such as interface name, MAC address, and index as well as wireless parameters (e.g., channel, transmission power, MAC level retries, and receiver sensitivity) for each virtual interface. The *REGISTRATION* message is sent by the VirtualMesh client tools just after start-up of the node. The *REGISTRATION* message is retransmitted if it is not acknowledged by the model through an *ACK* message within a predefined

4.3. VIRTUALMESH COMMUNICATION PROTOCOL

time-out, e.g., 10 s. After successful reception of the acknowledgement, the node can start transmitting its wireless traffic to the model. Upon node registration, the simulation model created an internal representation of the external node.

If a node leaves the wireless emulation, a *DE-REGISTRATION* message is sent to the simulation model. The simulation server then removes the node from the simulation model. After de-registration, the simulation server neither accepts traffic from this node nor does it forward traffic to the node.

DATA messages are used for encapsulating regular wireless network traffic and to exchange it between nodes and the simulation model. Network traffic is sent as *DATA* messages. A *DATA* message contains the intercepted Ethernet frame, its size, the message and host identification and the involved virtual interface. In VirtualMesh, the management network uses a larger maximum transfer unit (MTU) to compensate for the overhead of these *DATA* messages and to guarantee that no packet fragmentation is necessary when using the standard MTU for the data traffic coming from the nodes. The commonly used 1 Gbps infrastructure network supports MTU sizes up to 9000 bytes. At startup, the VirtualMesh client tools check the correctness of MTU sizes of the management network.

If any dynamic wireless parameters, such as the current communication channel and transmission power, have changed on the node, a *CONFIGURATION* message is sent to the model. The *CONFIGURATION* message includes the host identification, the index of the interface with the changes, and all the changed parameters of the interfaces as type/value tuples. After the parameters are supplied to the model, it can calculate the simulation behaviour. In order to minimise the overhead of message handling in the simulation model and to avoid blocking situations, VirtualMesh does not acknowledge *CONFIGURATION* messages. In the worst case, emulation continues using the old values.

In order to ensure the correct behaviour, a node using the VirtualMesh communication protocol can be in one of three protocol states. The node is either in the *unconnected*, *registration pending* or *connected* state. Being unconnected, the node has no connection to the simulation server. The node drops any wireless traffic. As soon as the node has sent a REGISTRATION message to the simulation server, it is in the *registration pending* state and waits for the confirmation of the registrations. The node still drops any wireless traffic. If no *ACK* has been received after five seconds, the node retransmits its *REGISTRATION* message. The registration is successful if the node receives an *ACK* message. The node is now *connected*. *DATA* and *CONFIGURATION* messages can now be exchanged between the node and the simulation server. By sending a *DE-REGISTRATION* message, the node leaves the wireless emulation and returns to the *unconnected* state.

The message flow between virtual/real WMN nodes and the simulation model is described in Section 4.5, whereas the message flow inside the simulation model is shown in Section 4.6.2. The complete protocols for node registration, de-registration, packet transmission, and configuration propagation are illustrated step-by-step in Section 4.6.3.

4.4 Host Virtualisation

Host virtualisation provides additional scalability for VirtualMesh. It reduces the necessary administrative effort for complex test scenarios and effectively lowers the costs of protocol development by reduced hardware cost and more flexibility in testing. Multiple virtual hosts, so-called guests, are running on top of a hypervisor or virtual machine monitor (VMM) on common x86-compatible hardware. The hypervisor presents a virtual hardware platform to the guests and monitors their execution. VirtualMesh can be used with different virtualisation products as long as they support Linux as guest operating system. For the VirtualMesh prototype, we have selected the XEN virtualisation platform [17]. XEN is a powerful and efficient open-source solution for host virtualisation. It can be integrated into several kernels of free operating systems such as Linux, NetBSD, FreeBSD, or OpenSolaris. It offers full virtualisation and para-virtualisation for node virtualisation.

Full virtualisation provides a complete simulation of the underlying hardware. A full-virtualised host uses the real device drivers, which then work on top of an emulated hardware layer. All software, including the operating system and the device drivers, run unmodified, in the same way as on the raw hardware. In contrast, para-virtualisation introduces some adaptations to the guest operating system. The software interface of a para-virtualised machine is similar, but not identical, to that of real hardware. Therefore, the drivers for network and block devices are replaced. In our scenario, the para-virtualised host employs our standard embedded Linux system, which is also running on a real node. It makes use of the new para-virtualisation feature of recent Linux kernels (paravirt_ops) that allows it to run on native hardware and as a para-virtualised machine. The para-virtualised operating system kernel accesses the network and block devices through a XEN specific driver. Since both approaches are available in Linux by default, they have been compared in Section 4.8.3 and the evaluation results have been discussed to motivate the choice of utilising para-virtualisation.

4.5 Client Implementation

Traffic interception/redirection at the MAC layer and emulation of the wireless medium represent the fundamental concept that VirtualMesh builds upon. In order to modify the network stack at the client, several tools are necessary for the

4.5. CLIENT IMPLEMENTATION

traffic interception/redirection. These VirtualMesh client tools are described in the following. Section 4.6 then describes the wireless simulation server.

The VirtualMesh client tools have to notify the simulation model about all packets that should be sent over the emulated wireless network. In our design the client tools simply forward the original packets to the *WlanModel* server. We introduce a virtual wireless interface, to which all applications can transparently send their traffic. Furthermore, the client nodes should be able to transparently configure the virtual wireless interfaces like normal wireless interfaces using the same API. Configuration changes are then automatically propagated to the simulation model during run-time. This offers enhanced flexibility in testing, e.g., of management architectures.

An important design choice is which operating system and mechanisms are used as a basis for the implementation of VirtualMesh. For prototyping VirtualMesh, the Linux operating system has been selected as it is widely used in embedded systems and in the research community. The open source character of Linux and the big developer community simplifies the implementation of extensions and modifications. Wireless device drivers as well as some external wireless driver projects, e.g., Madwifi [186] providing drivers for Atheros wireless cards, are accessed through the Wireless Extension (WE) [189] API. The WE is a set of commands that controls the kernel settings through the *ioctl* device. User-space tools such as wireless-tools [190] and *wpa_supplicant* [120] use the WE API to configure the wireless devices. Recently, a major kernel development step tries to unify the wireless support. Recent kernels, therefore, include a new Netlink-based interface (nl80211 [91]), which is shared by some newer wireless drivers. Moreover, a new wireless configuration tool *iw* [21] has been introduced. Migration to the new API is an on-going development.

Considering the different technologies currently available in the wireless support of Linux, VirtualMesh should provide transparent virtual interfaces, but still avoid being dependent on kernel structures due to on-going development. There are principally two possibilities to provide a transparent virtual wireless interface in Linux. The VirtualMesh functionality could be implemented as a new dedicated Linux kernel module. This kernel module would extend the packet encapsulation of the IP-in-IP device with the traditional WE API. Additionally, the new device would be configured by a new user-space utility communicating with the kernel, e.g., via Netlink [91] or sysfs [127]. A major advantage of this approach is that no alternation of existing tools (e.g., *ifconfig, iwconfig*) is necessary due to the same API (WE). Another advantage is a lower packet delay, as all performance critical parts reside directly within the kernel. However, there is an important drawback of this approach: any future change in Linux kernel wireless API may break the compatibility to the new module, as it is not part of the kernel source tree. As experimental network features are often used in embedded domain, it is likely that a new incom-

4.5. CLIENT IMPLEMENTATION

	Own driver in the Linux kernel	Existing TUN/TAP driver
Compatibility with existing tools	+ full	(+) inclusion of header file
WE API compatibility	+ full	(+) imitation of WE API
Dependency on Linux kernel development	- high	+ low, TUN/TAP updated as part of official kernel tree
Development effort	- high (kernel space)	+ low (user space)
Portability	- limited to Linux	(+) portable POSIX systems with TUN/TAP driver

Table 4.1: Possible solutions for the implementation of the virtual wireless interface.

patible kernel version may be required soon. The second approach is to integrate as much functionality as possible in user-space by using the existing TUN/TAP [109] device driver. The operating system considers a TUN/TAP interface as a normal network device. Instead of forwarding the received packets to a hardware device, a TUN/TAP interface forwards the packets to a user-space process. In user-space, the packets are then encapsulated and sent to the simulation server. The wireless device configuration can be handled by a shared library, which manages the internal device state. Obviously, the additional context switches during the packet forwarding introduce additional delays. Moreover, wireless tools need to be enhanced to query the external settings of virtual wireless interface. By imitating the WE API, compatibility can be still guaranteed. As this solution uses the standard Ethernet interface, it provides transparent network access for applications. Moreover, it simplifies the development, as the client tools are completely implemented in user-space. Table 4.1 summarises the discussion of both possible implementation approaches.

In favour of this simplified development, a broader applicability, and lower dependency on future kernel developments we decided against implementing a new kernel module and selected the second approach using the existing TUN/TAP module and user-space tools. In our approach, the testing environment is not bound to a specific out-of-tree kernel add-on and the daemon could even be ported to other operating system platforms that offer a POSIX system interface and a TUN/TAP driver (e.g., *BSD, Darwin). The remaining dependency is limited to the implementation of the current configuration interface.

Our virtual wireless device is built on top of the TUN/TAP device of the Linux kernel. The VirtualMesh client tools consist of three parts: the virtual interface

4.5. CLIENT IMPLEMENTATION

library *libvif*, the *vifctl* utility and the system service *iwconnect* (see Figure 4.4). The library *libvif* abstracts the access to the virtual interfaces including their states. It provides a WE compatible API to modify the virtual interface from existing configuration tools. This is achieved by imitating the *ioctl* system calls in *libvif* and a small modification in configuration tools to call the *libvif* functions if the kernel *ioctl* calls fail. Using the utility *vifctl*, virtual interfaces are created or deleted. The *iwconnect* system service connects the virtual wireless interface to the simulation server. It processes any Ethernet frames received from the TUN/TAP device and forwards the encapsulated frames to the simulation model. In the reverse direction, the *iwconnect* re-injects the network traffic that it receives from the simulation model back into the Linux networking stack via the TUN/TAP device.

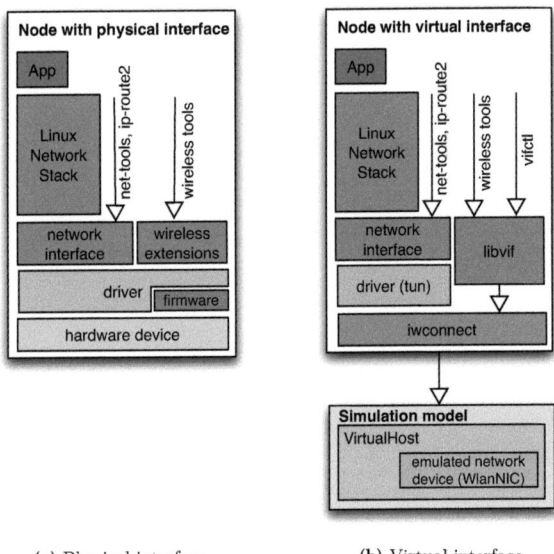

(a) Physical interface. (b) Virtual interface

Figure 4.4: A node with native Linux network stack (a) and a node with our virtual network interface (b) (*iwconnect*) communicating with the OMNeT++ simulation model.

In order to be operational, a network interface of a node has to be parametrised before its usage. It is configured during the machine installation and adapted during the network operation. A standard Linux network interface (see Figure 4.4a) is configured using net-tools or using the ip-route2 suite (i.e., using the commands *ifconfig*

4.5. CLIENT IMPLEMENTATION

or *ip*). Additionally, for wireless devices, wireless parameters such as wireless channel, operation mode, transmission power, Ready-to-Send/Clear-to-Send threshold, and encryption, are set by the wireless-tools (e.g., *iwconfig*) through the WE API of Linux. By using the kernel's TUN/TAP driver, our virtual device behaves the same as any Linux network device. Hence, no changes in the network configuration itself are required. Furthermore, the wireless parameters of our virtual interface can be set by a slightly patched version of wireless tools such as *iwconfig* (see Figure 4.4b), which then sets the parameters using *libvif*.

The virtual interface library (*libvif*) abstracts the access to the virtual wireless interfaces. Figure 4.5 illustrates this access to the virtual wireless interfaces. Instead of accessing parameters in the Linux kernel using the WE API, applications can modify the parameters of the virtual wireless interface through the shared library *libvif*. In contrast to normal device drivers, where the properties are stored in the kernel address space, the properties of the virtual interface are stored in a persistent global address space, from which they are loaded when the shared library *libvif* is accessed. Therefore, *libvif* offers a public API, which imitates the WE API of the kernel. Moreover, it provides *vifctl* device management functions to create and delete the interfaces. The shared library *libvif* is not directly involved in traffic redirection or wireless emulation. It just provides *iwconnect* the necessary information about the virtual interfaces for the packet forwarding to the simulation server.

The system service *iwconnect* is responsible for the communication with the simulation server. It receives all packets transmitted to the virtual interface and encapsulates them in new packets, which are sent to the host running the simulation model. In the opposite direction, *iwconnect* is listening on a UDP port for packets coming from the simulation model. These packets are then de-capsulated and original Ethernet frames are injected back into the network stack via the virtual interface, which then passes them to the application. Figure 4.6 illustrates this process. The numbers in the figure correspond to the individual steps taken. The complete wireless traffic of the node is processed by the virtual interface, the *iwconnect* system service, and the simulation model in the same manner.

Figure 4.6 shows the packet flow from the application at source node S to the destination node D. Both nodes are connected to the simulation model on host H. The application at node S sends the packets to the Linux network stack (1) where they are intercepted by the virtual wireless interface *vifx* (2). The original Ethernet frames are then redirected to *iwconnect* (3), which encapsulates them in new packets (4). These packets are transmitted through the Ethernet interface *ethx* (5) to the simulation model on host H (6). At host H, the packets are fed into the simulation model (see Section 4.6 for details). After processing in the simulation model, the resulting packets are encapsulated again and sent to their destination node D (7). There, the packets are received via the Ethernet interface *ethx* (8) and

4.5. CLIENT IMPLEMENTATION

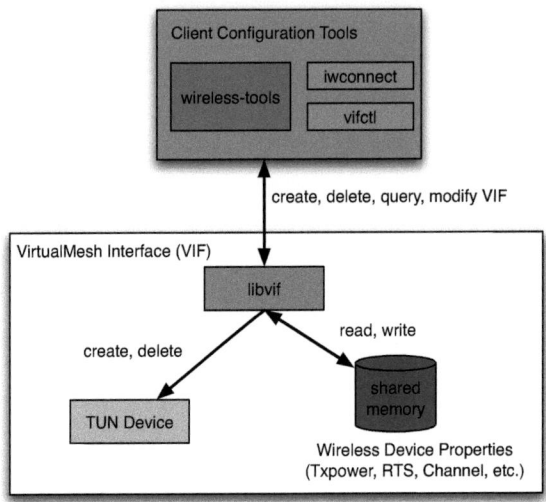

Figure 4.5: Access to virtual interfaces and its parameters using *libvif*.

the *iwconnect* system service (9). The *iwconnect* service extracts the packets and injects them back into the network stack via the virtual interface *vifx* (10). Finally, the application at node *D* receives the packets (11). This packet redirection is fully transparent for the applications and the network stack.

For accurate simulations, the simulation model needs to incorporate several additional static and dynamic parameters describing the external nodes and the current configurations of their wireless interfaces. Static parameters (e.g., IP address and listening port of the *iwconnect* system service) are set at start-up of the node. *iwconnect* has to register itself at the model by a *REGISTRATION* message (see Section 4.3). This message also includes the initial values of the dynamic wireless parameter such as channel and transmission power. During run-time, further configuration changes are propagated by *CONFIGURATION* messages to the simulation model.

The propagation of wireless parameters to the simulation model during run-time is a sophisticated feature that has been introduced by VirtualMesh and was not present in any other wireless emulation solution. The simulation model is automatically reconfigured with the wireless parameters that are set dynamically by the usual configuration tools at the wireless nodes. This offers possibilities for testing

4.5. CLIENT IMPLEMENTATION

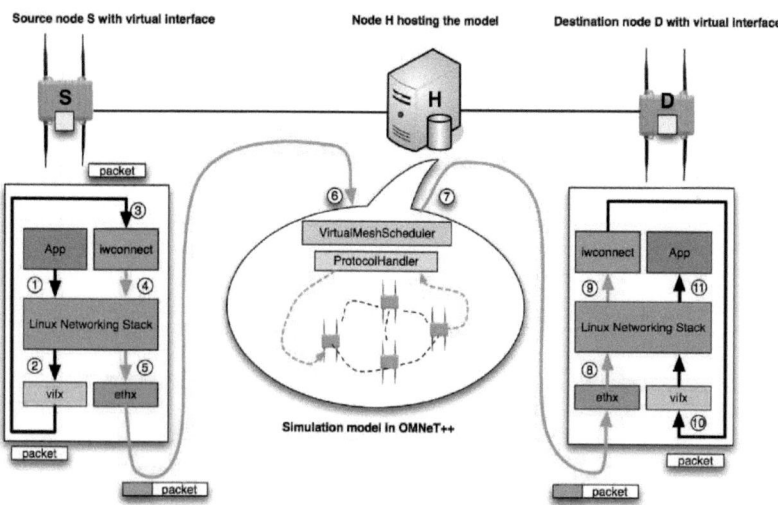

Figure 4.6: Packet flow between two nodes interconnected by the OMNeT++ simulation model.

management architectures that reconfigure the wireless interfaces on VirtualMesh. If VirtualMesh is extended with feedback mechanisms from the network emulation to the virtual wireless driver, even the multi-channel and multi-interface framework Net-X [51, 115] could be tested on top of VirtualMesh.

The propagation feature is implemented by standard POSIX message exchange based on inter-process communication (IPC). At node start-up, the system service *iwconnect* creates a message queue, which then is polled for notifications. If a configuration tool, e.g., iwconfig, or any other application modifies a parameter of the virtual wireless interface through the shared library *libvif*, the shared library sends a notification messages to the IPC queue. The system service *iwconnect* receives this notification and sends the changed parameters with a CONFIGURATION message to the simulation server, which then updates the simulation model. The notification mechanism can be extended to include also the dynamic management of wireless devices or even information about power management, which can then be reflected in the simulation model.

4.6 Wireless Simulation Server

The central part of VirtualMesh's wireless emulation is the wireless simulation server. It hosts the simulation model *WlanModel*, which processes the original link layer traffic received from the wireless nodes and models the emulated wireless medium. The *WlanModel* receives traffic coming from external nodes, calculates the system's response, and then sends the processed packets back to external nodes. The simulation model has been written for the network simulator OMNeT++ [195, 196]. In order to calculate the network response in the emulated wireless medium, the *WlanModel* employs the IEEE 802.11b implementation of the INET[94] simulation framework. VirtualMesh extends the existing simulation models to represent the involved wireless nodes within the simulation model using their virtual interface parameters. It further adds an own real-time scheduler to the simulation core of OMNeT++. This scheduler receives the forwarded wireless traffic from the nodes' *iwconnect* system service and injects it into the wireless simulation. The nodes can be freely positioned in the simulation model regardless of their physical position or virtual nature. Node mobility can be supported be either using existing mobility models of OMNeT++ or by the inclusion of mobility traces in the simulation model. In the following, individual components of the *WlanModel*, its packet flow, and the different processes for node registration, node de-registration, packet transmission, packet reception, and the configuration are discussed in more detail.

4.6.1 Components

The *WlanModel* has been implemented using and extending the OMNeT++ network simulator, which provides an advanced module system. In order to process the external traffic of the VirtualMesh nodes, the *WlanModel* adds a new real-time scheduler *VirtualMeshScheduler* to the OMNeT++ simulation core and provides a dynamic wireless network simulation model using several newly developed and some existing modules. The newly developed modules are the real-time scheduler *VirtualMeshScheduler*, the *ProtocolHandler*, *NodeManager*, and the *VirtualHost* module including the *VifBackend*. The *WlanModel* uses existing modules to model the wireless network behaviour. This includes the INET implementation of an IEEE 802.11b network stack *Ieee-80211NicAdHoc* (*WlanNIC*), the radio propagation model driven by INET's *ChannelControl*, and various mobility models.

The core component of the *WlanModel* is the new real-time scheduler *VirtualMeshScheduler* (see Figure 4.7). It replaces the standard scheduler of OMNeT++, which co-ordinates the event messages that are used for communication between the modules in OMNeT++. It is a soft real-time scheduler and adds the required functionality for VirtualMesh. The *VirtualMeshScheduler* includes the external traffic received from the wireless nodes in the simulation and adds messages to

4.6. WIRELESS SIMULATION SERVER

the event scheduling queue of OMNeT++. Therefore, it listens on the port 2424 for arriving packets. VirtualMesh's soft real-time scheduler uses the following scheduling policy: Internal messages and arriving forwarded network packets are added to a global message queue with an event time. The scheduler checks the queue for events to be processed. If there are currently no messages with due event times, the scheduler listens on the network socket for further arriving packets. If a network packet has arrived, the scheduler immediately notifies the *ProtocolHandler* module, which then takes the care of the packet. After a time-out, the scheduler then checks the global queue again for the next iteration.

The *ProtocolHandler* module handles the communication with the external real or virtualised wireless nodes and implements the processing of messages of the VirtualMesh communication protocol (see Section 4.3). The *ProtocolHandler* module forwards the information to the *NodeManager* or a *VirtualHost* module depending on the received message type (*REGISTRATION, DE-REGISTRATION, CONFIGURATION,* or *DATA*).

The *NodeManager* module manages the wireless nodes that are participating in the wireless simulation. Nodes are dynamically registered and de-registered at the *NodeManager* through the *iwconnect* system service running on each node. Upon reception of a *REGISTRATION* message the *NodeManager* creates and registers a new *VirtualHost*. It removes the *VirtualHost* after reception of a *DE-REGISTRATION* message.

The *VirtualHost* module represents a wireless node and all its parameters within the simulation. It is a compound module consisting of the modules *VifBackend*, *WlanNIC* and some other INET modules. It stores the parameters of the corresponding wireless node such as the host name, the host identification, and host address. The node's current position is handled by an INET mobility model that also supports the configuration of various mobility models.

The *VifBackend* module handles the communication between the interface representation in the simulation model and the virtual wireless interface of the corresponding node. Upon reception of a *DATA* message, *VifBackend* creates a new *RAWEtherFrame* with the Ethernet frame encapsulated in the *DATA* message. The *RAWEtherFrame* simply wraps the original Ethernet frame within the simulation. Upon reception of a *RAWEtherFrame* from the simulation model, the *VifBackend* send the Ethernet frame as *DATA* message to the external node. Upon reception of a *CONFIGURATION*, it adapts the wireless settings of the *WlanNIC*.

The *WlanNIC*, i.e., *Ieee-80211NicAdHoc*, of the INET framework, implements an IEEE 802.11 wireless network adapter. It models the IEEE 802.11 MAC layer and implements the radio propagation model. It supports the wireless ad-hoc communication necessary for WMNs. Its wireless channel can be modified during the simulation, which is an essential feature for the dynamic configuration propagation

4.6. WIRELESS SIMULATION SERVER

in VirtualMesh.

The *ChannelControl* is the central component of the INET wireless network model. It monitors which nodes are within each others' communication range. These nodes receive the frame and decide by themselves if they can receive the packet or have to discard it. The *ChannelControl* supports multiple channels, but does not model either co-channel or adjacent channel interference.

4.6.2 Message Flow

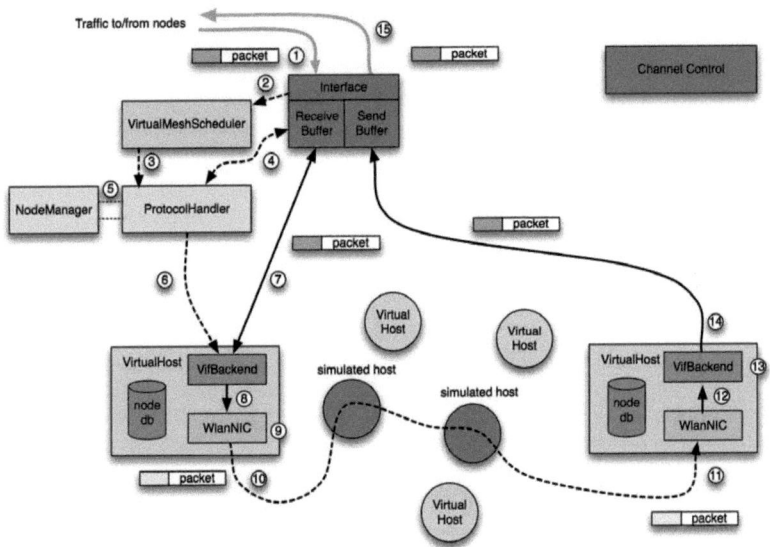

Figure 4.7: Message flow inside the simulation model 'WlanModel'.

The message flow within the *WlanModel* is shown in Figure 4.7. It shows the individual components, which are described in more detail in the following. The numbers in the brackets represent the steps in Figure 4.7.

Upon packet reception (1), the new packet is stored in the interface receive buffer, the *VirtualMeshScheduler* is informed (2) and schedules a notification message with the reception time for the *ProtocolHandler* module (3) in the global message queue of the simulation. This message queue is then processed step-by-step, ensuring that required timing constraints are met.

4.6. WIRELESS SIMULATION SERVER

As soon as the *ProtocolHandler* module gets a notification message (3), it processes the received messages coming from outside the simulator (4). If a new external node registers its presence to the model, the *ProtocolHandler* calls the *NodeManager* (5), which handles the *REGISTRATION* message. The *NodeManager* is responsible for the administration of external nodes inside the simulation model. If the registering node does not already exist in the simulation model, the *NodeManager* creates a new instance of *VirtualHost* and saves all node attributes (host name, infrastructure IP address, listening port) to the node database of the *VirtualHost*. After successful initialisation, the *VirtualHost* acknowledges its presence to the external node by an *ACK* message. If the node was already existing, the *NodeManager* only instructs the *VirtualHost* to acknowledge its presence again.

Upon *DATA* message reception (4), the *ProtocolHandler* first checks whether the sending node has already registered at the *NodeManager* (5). If no registration exists, the packet is dropped immediately. Otherwise, the *ProtocolHandler* notifies the *VifBackend* of the corresponding *VirtualHost* (6). The *VifBackend* processes the packet (7). The encapsulated original Ethernet frame is included in a new *RAWEtherFrame* packet, which is then transmitted to the *WlanNIC* of the *VirtualHost* (8). The *WlanNIC* uses the current wireless parameters for the transmission to the next node (9). Changes of the wireless parameters of external nodes are propagated to the simulation model by the transmission of a *CONFIGURATION* message. The *ProtocolHandler* is only involved in the processing of incoming traffic to the simulator. Henceforth, the existing IEEE 802.11 model implementations of INET takes care of the packet (10) until it has been received again by a *VirtualHost* module (11). The *WlanNIC* checks whether the packet belongs to this node by checking the MAC addresses. If yes, it is forwarded to the *MsgHandler* (12). Otherwise, it is omitted. The *VifBackend* generates a new *DATA* message that includes the packet (13). When the *Interface Send Buffer* receives a *DATA* packet from the *VifBackend* (14), the packet is finally forwarded over the system network (15) to the external node.

4.6.3 Protocols

In the following, the different processes in VirtualMesh are shown step-by-step, covering the communication between the real node and simulation model as well as the communication inside the simulation model. Five processes exist in VirtualMesh. First, the external node has to register itself at the simulation model (node registration). It can also cancel its registration within the model afterwards (node deregistration). When successfully registered, the external node can transmit packets (packet transmission) to its representation in the simulation model. After packet processing inside the simulated network, an internal representation of a node receives the packet and then transmits it to the connected external node (packet reception).

4.6. WIRELESS SIMULATION SERVER

By the transmission of a *CONFIGURATION* message, the external nodes can modify their interface parameters (node configuration). The numbers in the brackets (4.6.x and 4.7.y) reflect the steps in Figure 4.6 and Figure 4.7. In addition, node registration, de-registration, and configuration are illustrated in Figure 4.8-4.10.

Node registration

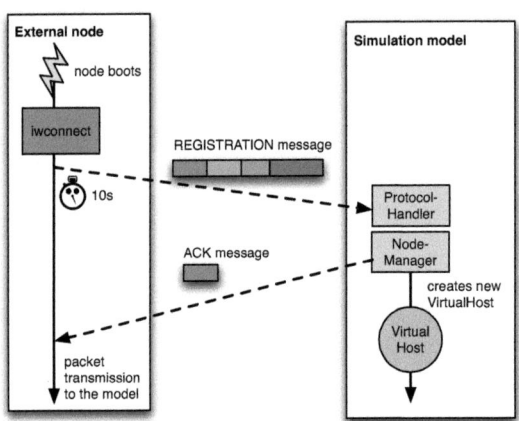

Figure 4.8: Node registration.

Figure 4.8 illustrates the node registration. A node is added to the simulation model after sending a *REGISTRATION* message. The detailed steps are as follows:

1. The node with a VirtualMesh interface boots. The configuration of the virtual interface contains the IP address and the port of the simulation model.

2. The node's *iwconnect* system service sends a *REGISTRATION* message to the model (4.6.4 - 4.6.6).

3. The simulation model adds the node to the *NodeManager* (4.7.1 - 4.7.5) and replies with an acknowledgement.

4. The *NodeManager* creates a new *VirtualHost*. The node database of the *VirtualHost* is initialised with the values of the *REGISTRATION* message. This includes host name, the infrastructure IP address and the listening port of the

4.6. WIRELESS SIMULATION SERVER

PacketModeller, number of interfaces and their configuration (name, MAC, transmission power, MAC level retries, receiver sensitivity etc.).

5. Positions and mobility patterns of the *VirtualHost* have to be configured in-advance by the OMNeT++ setup file and are, therefore, already present inside the simulation model.

6. Upon reception of the acknowledgement, the node is registered and can send/receive traffic to/from the simulation model.

Node de-registration

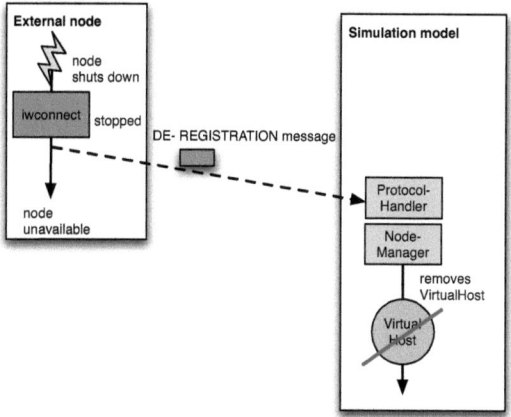

Figure 4.9: Node de-registration.

In Figure 4.9, a node de-registers by sending a *DE-REGISTRATION* message to the simulation model. The node is then removed from the simulation model. The detailed steps are as follows:

1. A shutdown of the *iwconnect* system service, e.g., when rebooting the external node, triggers the transmission of a *DE-REGISTRATION* message. It forces the *VirtualHost* to leave the emulated network (4.6.4 - 4.6.6).

2. Upon reception of a *DE-REGISTRATION* message, the *ProtocolHandler* invokes a node removal by the *NodeManager* (4.7.1 - 4.7.5). In order to avoid

4.6. WIRELESS SIMULATION SERVER

delays and blocking when rebooting the external node, the reception of a *DE-REGISTRATION* message is not acknowledged.

3. The *NodeManager* removes the corresponding *VirtualHost*.

Node registration and de-registration allow the emulation of dynamic networks. Nodes can join and leave the network. They are automatically added or removed from the simulation model. This is beneficial, for example, to test a configuration and management framework. The effects of nodes rebooting, or becoming unavailable, at certain configuration times can be evaluated.

Packet transmission

The process for packet transmission from a node with a virtual wireless interface to the emulation model is as follows:

1. The source application at the node sends a packet to the virtual interface *vif* (4.6.1, 4.6.2).

2. The *iwconnect* system service encapsulates this packet and then redirects it as a *DATA* message to the simulation model (4.6.3 - 4.6.6).

3. The *DATA* message is received by the simulation model and it is stored in the *Receive Buffer*(4.7.1).

4. The *VirtualMeshScheduler* is notified and schedules the notification message to the *ProtocolHandler* (4.7.2, 4.7.3).

5. The *ProtocolHandler* checks message type and, via *NodeManager*, whether the sender of the *DATA* message exists (4.7.4, 4.7.5). If it exists, it receives a pointer to the corresponding *VirtualHost*.

6. The *ProtocolHandler* calls the *VifBackend* of the *VirtualHost* to handle the packet (4.7.6).

7. The *VifBackend* gets the *DATA* message and a new *RAWEtherFrame* is transmitted (4.7.7 - 4.7.10).

    (a) The *DATA* message is read from the *Receive Buffer* (4.7.7).

    (b) A new *RAWEtherFrame* packet is created (4.7.8).

    (c) The original Ethernet frame of the *DATA* message is copied to the new packet.

4.6. WIRELESS SIMULATION SERVER

(d) The destination address of the *RAWEtherFrame* packet is set to the destination MAC address of the original Ethernet frame.

(e) The packet is passed to the *WlanNIC* (4.7.9).

(f) The packet is transmitted inside the simulation model (4.7.10).

(g) The *RAWEtherFrame* message is transmitted to the corresponding *VirtualHost* (4.7.9, 4.7.10).

Packet reception

The following steps are necessary for the packet reception by a node with a virtual wireless interface:

1. *VirtualHost* receives a packet through the *WlanNIC* (4.7.11) and passes it to the *VifBackend* (4.7.12).

2. The *VifBackend* encapsulates the Ethernet frame inside a new *DATA* message, including the infrastructure IP address and listening port of the corresponding external node, and then transmits it to the *Interface Send Buffer* of the model (4.7.13, 4.7.14).

3. The *DATA* message is then forwarded to the external node (4.7.15).

4. The *iwconnect* system service at the external node then decapsulates the packet and injects the Ethernet packet to the network stack of the node (via the virtual interface *vifx*) (4.6.7 - 4.6.11).

Node configuration

In order to propagate modified wireless parameters to the simulation model, a *CONFIGURATION* message is transmitted from the node to the simulation model (see Figure 4.10). The detailed steps of this process are explained in the following:

1. The wireless interface configuration of the external node is modified by the wireless tools such as the *iwconfig* utility.

2. The patched version of the *iwconfig* tool propagates all wireless configuration changes via *libvif* to the shared memory *vif* configuration.

3. Then, the *iwconnect* system service is triggered to send a *CONFIGURATION* message to the simulation model (4.6.3 - 4.6.6).

4. The simulation model then processes the *CONFIGURATION* messages (4.7.1 - 4.7.7).

4.7. ADAM/VIRTUALMESH INTEGRATION

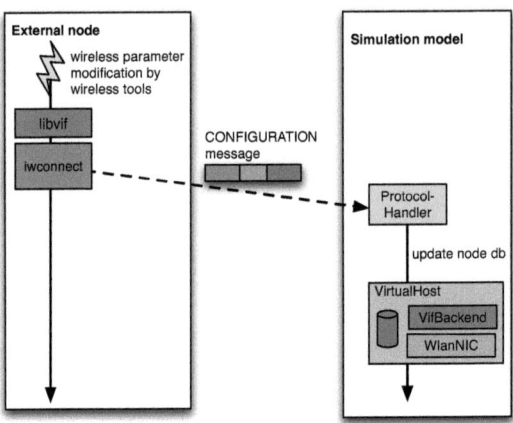

Figure 4.10: Node configuration.

(a) The *ProtocolHandler* reads the *CONFIGURATION* message (4.7.4).

(b) It then checks whether a representation of the external node exists in the model with the help of the *NodeManager* (4.7.5).

(c) The *ProtocolHandler* then invokes the *VifBackend* of the corresponding *VirtualHost* to process the *CONFIGURATION* message (4.7.6, 4.7.7).

(d) The *VifBackend* stores the new wireless parameter values in the node database.

(e) Henceforth, the *VirtualHost* uses the new values for packet transmissions inside the model.

The nodes can modify the parameters of their wireless network interfaces by the wireless tools (*iwconfig* utility), at any time during the simulation. Even scenarios with dynamic multi-channel communication are possible. However, the simulation model has to be extended to support them, principally the used network simulator OMNeT++ already supports them.

4.7 ADAM/VirtualMesh Integration

VirtualMesh provides an ideal testing framework for ADAM (see Chapter 3) with its distributed software and configuration update mechanism based on *cfengine* (see

Section 3.4.1).

In order to run an ADAM image in a virtualised environment, some modifications were necessary to the Linux distribution produced by ADAM. We added a new node profile *'xen'*, which compromises of all necessary modifications to ADAM's *build-tool*. These modifications include an adapted configuration of the Linux kernel to support XEN virtualisation and the TUN device driver, the addition of the VirtualMesh client tools, and a new network initialisation script. The network initialisation script creates virtual wireless interfaces with *vifctl*, in case no physical wireless interfaces are found at start-up. This guarantees that the same ADAM image can be run on real node hardware and on virtualised XEN nodes within VirtualMesh. ADAM's *image-tool* has been extended to create virtual disk images with the boot loader Grub that are suitable for XEN. These modifications guarantee full compatibility of ADAM with both the VirtualMesh XEN environment and the simulation server.

Due to XEN's boot loader pygrub [106], which behaves like Grub and reads the Grub configuration, ADAM's safe update procedure works without any further modification in a VirtualMesh emulation. Thus, network configuration and software images can be updated over the emulated network. Further developments of ADAM can be fully tested with VirtualMesh.

4.8 Evaluation

VirtualMesh has been evaluated considering functional / qualitative and performance / quantitative aspects. VirtualMesh is using an emulation approach to create a testing infrastructure for real implementations of protocols and architectures for wireless mesh and ad-hoc networks under various, but realistic conditions. It supports the participation of real Linux based hosts as wireless nodes in the emulated network. In order to be usable for its main purpose, i.e., pre-deployment tests with the real software stack, VirtualMesh has to fulfil the following requirements:

- VirtualMesh should not change the behaviour of the network protocols and applications running on top of it. It has to be fully transparent for the upper layers. This has been verified in our functional evaluation in Section 4.8.2.

- In addition, neither bandwidth nor delays of the wireless network should be heavily influenced by VirtualMesh. Due to traffic interception, traffic redirection to a simulation model, and the optional node virtualisation, the architecture of VirtualMesh introduces some additional delays to the system. The performance evaluation in Section 4.8.3 quantifies the effect of VirtualMesh on delays and network bandwidth.

The distributed VirtualMesh architecture is very flexible: a VirtualMesh system can simply be extended by additional nodes or virtualisation hosts connected to

4.8. EVALUATION

the infrastructure network. Unfortunately, the infrastructure network used in the distributed VirtualMesh architecture also introduces an additional small delay to all wireless communication (see Section 4.8.3). In addition, the infrastructure network limits the entire bandwidth of all traffic processed by VirtualMesh. A usual setup for VirtualMesh uses 1 Gbps Ethernet as network technology. Employing the low latency interconnection such as Infiniband [56] may further reduce the transmission delay. However, the packet processing of the OMNeT++ simulation model primarily determines the additional delay for a wireless packet transmission in VirtualMesh. The computations of the simulation models may be too intensive to be performed in real-time, which might add some small delays not found in a real wireless transmission. Therefore, VirtualMesh cannot provide any hard real-time guarantee for the wireless emulation. Nevertheless, the accuracy of the soft real-time scheduler is sufficient for most scenarios.

Setting up a test scenario using VirtualMesh is more demanding than a pure simulation and requires some additional work to be done by the user/developer. VirtualMesh's approach with multiple real nodes that participate in an emulated wireless network adds some additional complexity to the setup of experiments and the data acquisition compared to a pure network simulation. In contrast to the autonomous self-contained system of a network simulation, VirtualMesh has a distributed concept to incorporate real wireless nodes as well as virtualised nodes in the wireless emulation. All applications communicating over the emulated wireless network are running on the different hosts. This requires event triggering and logging to be placed directly on the individual nodes. If the execution of all test events and the data acquisition is automated by scripts, VirtualMesh still achieves a high degree of repeatability even when using complex scenarios. These scripts also allow repeating the scenario with different parameter sets. As result of these efforts, the measurements made within VirtualMesh mirror the normal operation of the nodes running a Linux-based operation system and the real applications with a high degree of accuracy and in a realistic scenario. This includes slight variations of the results that may be observed for test runs made under the same conditions.

In the following sections, VirtualMesh has been evaluated under functional and performance aspects. First, the general experimental setup is described in Section 4.8.1. Section 4.8.2 then provides a functional evaluation of VirtualMesh, whereas Section 4.8.3 evaluates the system performance of VirtualMesh under different conditions.

4.8.1 VirtualMesh Test Setup

The experimental setup for the VirtualMesh evaluation is shown in Figure 4.11. It consists of two servers connected over a 1 Gbps Ethernet cross-link, dedicated solely to the infrastructure traffic of VirtualMesh and has an MTU set to 2000

4.8. EVALUATION

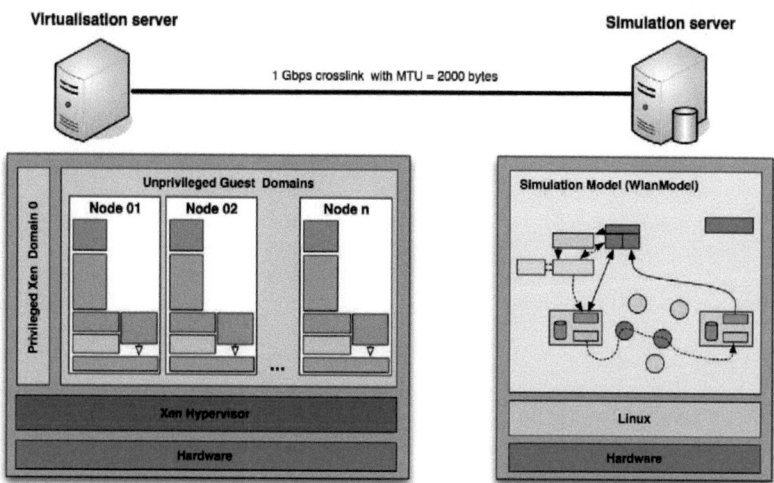

Figure 4.11: Experimental setup with multiple virtualised wireless nodes running on a XEN virtualisation server and a simulation server hosting the WlanModel.

bytes. The first server runs XEN virtualisation to host multiple virtual machines, which represent the wireless nodes in our setup. The virtual machines are running the embedded Linux distribution built by the ADAM framework (see Chapter 3) and include the VirtualMesh client tools. The second server runs the simulation model *WlanModel* in command-line mode as a prioritised system service on a Linux operation system. In order to reduce system latencies, we granted the highest CPU scheduling policy to the *WlanModel* process.

The used configuration of the *WlanModel* emulates an IEEE 802.11b based wireless network. All wireless parameters are summarised in Table 4.2. They are divided into static and dynamic parameters. Static simulation parameters are pre-configured before starting the wireless emulation. Dynamic parameters are configured directly at the wireless nodes. They are constantly reflected in the simulation model after the nodes have been registered at the simulation model. They can be dynamically adapted. For example, the dynamic driver reconfiguration can provide topology control by adapting the transmission power of the nodes.

4.8. EVALUATION

Description	Parameter Type	Value
Number of radio channels	static	13
Maximum transmission power	static	50.0 mW
Signal attenuation threshold	static	-110 dBm
Path loss exponent α	static	2
Radio carrier frequency	static	2.4 GHz
Wireless device bit rate	static	11 Mbps
Contention window for normal data frames	static	32 packets
Contention window for broadcast frames	static	32 packets
Maximum queue length in frames	static	14 packets
Base (thermal) noise level	static	-110 dBm
Signal/Noise ratio threshold (SNR)	static	4 dB
Wireless channel	dynamic	1
Transmission power	dynamic	17 dBm
Radio sensitivity	dynamic	-85 mW
RTS/CTS threshold	dynamic	2346 B (off)
Maximum number of retries	dynamic	7

Table 4.2: VirtualMesh wireless configuration settings consisting of static parameters directly set in the simulation model *WlanModel* and dynamic parameters propagated from the virtual interfaces.

4.8.2 Functional Evaluation using ADAM

VirtualMesh has been designed to integrate real communication nodes in an emulated wireless network. We implemented the VirtualMesh client-tools for mesh nodes running a Linux operating system. Due to modest soft- and hardware requirements, VirtualMesh runs on most Linux installations including our own embedded Linux distribution ADAM (see Chapter 3). This section provides a functional evaluation of VirtualMesh using ADAM in the XEN virtualisation setup shown in Figure 4.11. We verified that VirtualMesh fulfils the following three functional requirements:

- Transparency for applications on standard Linux clients
- Transparency for virtualised nodes running our ADAM Linux distribution
- Proper operation of a routing protocol running on virtually mobile nodes

First functional tests are performed with two physical machines running the Ubuntu Linux distribution and the VirtualMesh client tools. The transparent operation of several existing Linux applications, such as remote administration with secure remote shell (ssh), file transfers using the file transfer protocol (FTP) and secure copy (scp), has been successfully verified in VirtualMesh.

4.8. EVALUATION

We repeated the basic tests using ADAM nodes to verify the transparent operation of VirtualMesh with ADAM. As expected, the network traffic is handled completely transparently. All services on the nodes can be accessed, e.g., the web server and the secure remote shell. The full protocol stack operates as being located in a real wireless network. This includes ARP, IPv4, IPv6, UDP, TCP, etc. Even ADAM's distributed software and configuration update mechanism based on *cfengine* (see Section 3.4.1) works just like in a real network. Network configuration and software images can be updated over the emulated network. Although the system update involves a reboot of the node, it can be performed without any problem due to the node registration and de-registration process of VirtualMesh. During the system update, the new image consisting of a new Linux kernel and a new root file system is copied to the node's virtual hard disk. After modification of the boot loader files, the ADAM node is rebooted using the updated system image. The system update works completely transparent in a VirtualMesh setup. This makes VirtualMesh a valuable tool for the development of ADAM management extensions.

As a last functional test, we have set up an ad-hoc routing scenario with twelve mobile wireless nodes running ADAM Linux. We configured OLSR as ad-hoc routing protocol on the nodes. The node mobility is pre-defined by a mobility trace. Some of the nodes are permanently moving. OLSR correctly adapts the routing tables according to the changing topologies. As OLSR runs as application on the nodes and all its control traffic is transferred through the virtual wireless device driver of VirtualMesh, OLSR operated fully transparent within the VirtualMesh emulated wireless scenario. OLSR cannot distinguish the emulated network compared from a real network. Therefore, VirtualMesh represents a valid tool for the evaluation of real implementations of layer-3 routing protocols, e.g., AODV, DSR, DSDV, TBRPF.

In contrast to layer-3 routing protocols, VirtualMesh cannot be used for testing the real implementations of layer-2 routing protocols, such as IEEE 802.11s, due to its current implementation with traffic redirection by a virtual wireless interface driver. For example, if a user wants to set up a scenario with IEEE 802.11s, he/she has to include the IEEE 802.11s functionality in the VirtualMesh simulation model.

Another limitation of VirtualMesh is that current implementation does not provide feedback of MAC and radio parameters from the simulation model to the virtual device driver. For example, if a cross-layer protocol requires the number of MAC level retransmissions, the current implementation of VirtualMesh does not delivers this parameter. VirtualMesh needs to be extended by a feedback mechanism before such cross-layer protocols could be tested completely in a VirtualMesh emulated network.

Our functional tests showed that VirtualMesh does not affect the normal operation of the participating nodes and applications. It is even able to imitate complex scenarios with software and configuration updates on the individual nodes. It pro-

4.8. EVALUATION

vides a full virtualisation of various wireless network scenarios without losing functionality of the nodes under test. In Section 4.8.3, the performance of VirtualMesh is evaluated and its overhead is quantified.

4.8.3 Performance Evaluation

In order to evaluate the performance of VirtualMesh, latency and throughput measurements have been performed. First, we present our latency measurements and quantify the effect of the infrastructure network, the virtualisation, and the wireless emulation. Second, the effect on achievable throughput is quantified.

Network Latency Measurements

Network latency usually is measured as round-trip time (RTT). It represents the time required to send a packet to another network host and to receive its immediate reply. In our evaluations, we used the standard *ping* tool from the *iputils* package [112]. It sends an ICMP Echo Request to the other network hosts and then measures the time until it receives the corresponding ICMP Echo Reply.

The RTT measurements were performed with different payload sizes. The payload sizes range from 56 bytes to 1472 bytes in the wireless emulation. In the infrastructure network, the evaluation includes payload sizes up to 1532 bytes in order to include the encapsulated packet with an MTU of 1500 bytes and 32 bytes for the VirtualMesh header. The measurement interval was varied from 0.1 to 1 second. All RTT measurements were run for 1'000 seconds. As the first RTT measurement in a sequence is sometimes significantly higher than the following ones due to the ARP look-up. Hence, it is omitted in our evaluation.

Infrastructure Network Latency

The VirtualMesh evaluation starts with the infrastructure network latency. The infrastructure network introduces three additional delaying effects for traffic transmitted in VirtualMesh. First, the transmission over the infrastructure network adds some delay. Each packet has been transmitted twice over the infrastructure network for each transmission over one emulated wireless link. Second, the delay may be further increased through system virtualisation. Third, the traffic en-decapsulation of the VirtualMesh communication protocol through *iwconnect* and the *WlanModel* adds some delay. The three delaying effects have been tested with the following host combinations:

(a) Two physical hosts connected via 1 Gbps cross-link

(b) Physical host to paravirtualised host connected via 1 Gbps cross-link

4.8. EVALUATION

(c) Physical host to full-virtualised host connected via 1 Gbps cross-link

(d) Two physical hosts connected via 1 Gbps cross-link and using the VirtualMesh communication protocol through *iwconnect*

(e) Physical host to paravirtualised host connected via 1 Gbps cross-link and using the VirtualMesh communication protocol through *iwconnect*

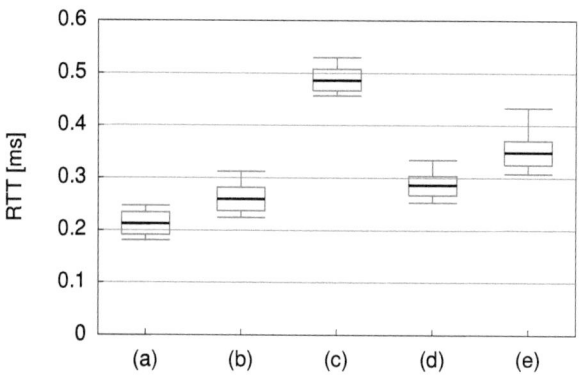

Figure 4.12: Summarised RTT results for quantifying infrastructure network delay.

Figure 4.12 shows the results for the five different scenarios. It combines the medians of all results received for the different packet sizes and data rates. It shows the maximal RTT as top bar, the minimal RTT with as bottom bar, the median as thick line in the middle, and the 25% and 75% quantiles of all measurement series. In all scenarios, the RTT increases for higher packet sizes in all scenarios, e.g., in (a) the median RTT is about 0.19 ms for 56 bytes payload and about 0.25 ms for 1532 bytes payload. The maximum RTT in Figure 4.12 depicts the average value for the maximum payload size.

As expected, the results for the virtualisation setups (b)(c) show some additional delays compared to the native scenario (a) due to the packet handling at the virtualisation server. When using paravirtualisation (b), the average additional delay is almost negligible with values between 0.04 to 0.06 ms. In contrast, the full virtualisation setup (c), using the hardware emulation layer (Hardware Virtual Machine), adds a significantly higher delay of up to 0.3 ms. Full virtualisation is, therefore, no option for VirtualMesh as it would have a huge impact on the emulation accuracy. As consequence, full virtualisation is completely omitted in the following evaluations.

4.8. EVALUATION

An interesting point is that the involvement of virtualisation in (b) and (c) does not modify the traffic characteristics, as the average standard deviation remains around 15 µs. When running the RTT measurements over the virtual interface using *iwconnect* and the VirtualMesh communication protocol over either a native link (d) or a physical-paravirtualised link (e), an average delay of 0.06 ms is added respectively to the delay of the native link (a) and the physical-paravirtualised link (b). Paravirtualisation further adds some additional delay in scenario (e), as the user-space daemon *iwconnect* is more sensitive to process scheduling of the hypervisor.

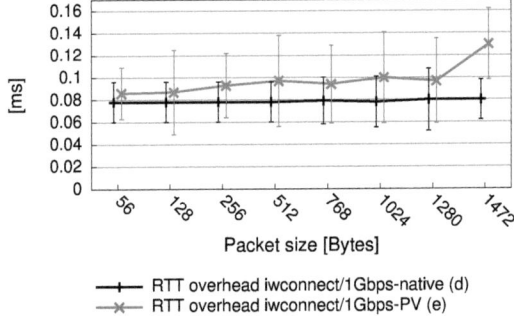

Figure 4.13: *iwconnect*/VirtualMesh communication protocol RTT overhead with respect to payload size.

Figures 4.13 and 4.14 further illustrate this effect. They compare the RTT overhead and the standard deviations of native and paravirtualised scenarios. Whereas the latency overhead remains constant with 80 µs over all payload sizes for the connection with two physical hosts, it slightly increases for the paravirtualised host with the payload size. The overhead for the connection with a paravirtualised host is about 20 µs higher than for connection with only physical hosts. In general, the standard deviation increases as soon as the traffic is forwarded by *iwconnect* over the VirtualMesh communication protocol. The results in Figure 4.14 show that shorter transmission intervals increase the amount of outliers and lead to a higher standard deviation in the paravirtualisation scenario.

In summary, both the paravirtualisation and the traffic redirection by *iwconnect* individually add additional delays in the range of 40 to 80 µs to the packet transmission in VirtualMesh. As full virtualisation in XEN multiplies the delays, it is not considered a reasonable option for VirtualMesh. When used in combination with virtualisation, the traffic redirection by *iwconnect* slightly modifies the traffic characteristics, which is shown by a higher standard deviation.

4.8. EVALUATION

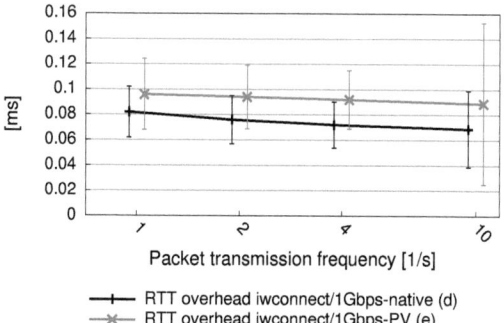

Figure 4.14: *iwconnect*/emulation protocol RTT overhead with respect to transmission interval.

Wireless Emulation Accuracy

After having quantified the delay introduced by the infrastructure network, the wireless emulation accuracy of the entire VirtualMesh system is now discussed. For evaluation, the measurements received with VirtualMesh are compared with a simple OMNeT++ simulation. First, the RTT over different distances between two ADAM wireless nodes connected to the simulation model are tested. The nodes have been placed in the simulation with distances of 1 m and 580 m. 580 m is the maximum transmission range that can be reached with the settings in Table 4.2. The theoretical difference of the two delays due to signal propagation is calculated with 3.86 μs. Pure simulation matches this result with a RTT of 1.242 ms for 1 m distance and 1.246 ms for 580 m. Emulation results in 1.6 ms RTT, which is about 0.35 ms higher than the values of pure simulation and has a higher standard deviation. It shows that VirtualMesh cannot realistically model delaying effects of only a few microseconds such as the influence of the distance on the propagation delay.

The next experiment evaluates the performance of VirtualMesh with different packet rates and payload sizes. The scenario consists of two ADAM wireless nodes with a distance of 300 m between them in the *WlanModel*. As no abnormalities have been observed in the comparison with different packet rates, only the RTT results for 1 packet/s and different payload sizes is shown in Figure 4.15. The complete result set can be found in [77]. The emulation results closely follow the simulation results, just with the previously discussed latency overhead of about 0.35 ms. The standard deviation of the VirtualMesh results also follows the measured standard deviation of the infrastructure network. The simulation results show a higher standard deviation,

4.8. EVALUATION

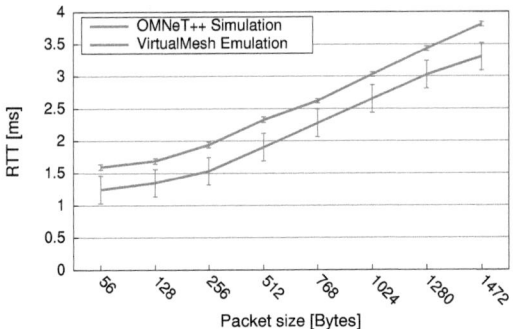

Figure 4.15: RTT with various payload sizes (distance = 300m, transmission interval = 1s).

which seems to be an anomaly in the simulation configuration or implementation.

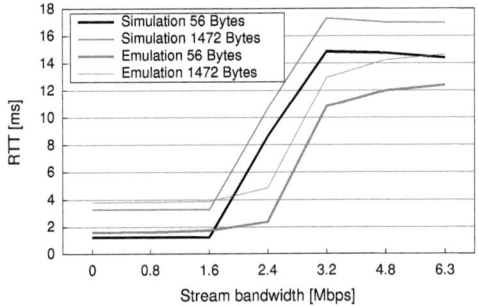

Figure 4.16: RTT with concurrent streams (distance = 300m, transmission interval = 1s).

In order to test VirtualMesh under load, a concurrent network stream has been added to the scenario described above. The concurrent TCP network stream should keep VirtualMesh busy. We evaluated stream bandwidth between 0.8 and 6.3 Mbps, whereas 4 Mbps starts to saturate the IEEE 802.11b wireless link. The TCP streams are produced by the *TCPSessionApp* for the simulation scenario and with *netcat* and *curl* for the VirtualMesh scenario. Figure 4.16 shows an exciting result. The results for VirtualMesh are additionally delayed by 0.35 ms compared to pure simulation, as measured previously. However, with a concurrent stream using more than 1.6 Mbps bandwidth, the RTT results for VirtualMesh are below the ones for the

4.8. EVALUATION

simulation. A possible explanation is the different implementation of traffic generation for simulation and VirtualMesh. Nevertheless, there is no negative influence of the concurrent stream to VirtualMesh. Moreover, the simulation suffers from an inaccurate handling of the concurrent stream as the RTT slightly decreases for bandwidth above 3.2 Mbps. VirtualMesh then displays its advantage, i.e., using real applications and network stacks, and produces results that matches the real world behaviour more closely than the simulation. This basic test shows the advantage of having real software and network stacks under test when using VirtualMesh's wireless emulation approach.

Simulation Latency Overhead

The scalability of VirtualMesh was tested in scenarios with additional wireless nodes. First, the influence of additional nodes that are not involved in the transmission is quantified. Due to position and state checks, any additional node increases the computational overhead of the propagation model, i.e., simulation model. In the test scenario, the RTT between two wireless nodes is tested with an increasing number of nearby nodes. The additional nodes are within transmission range and placed in a grid around the two communication peers. The influence of uninvolved nodes on the RTT measurements is shown in Figure 4.17. The overhead is roughly proportional to the number of additional nodes. In this scenario, VirtualMesh suffers from the soft real-time scheduling used in the simulation model. Virtualisation does not influence the result. The additional delays are only caused by calculations within the *WlanModel*. The *WlanModel* verifies for each node if it could receive the currently transmitted packet. The required calculations slightly delay the packet transmission in VirtualMesh for increased number of nodes.

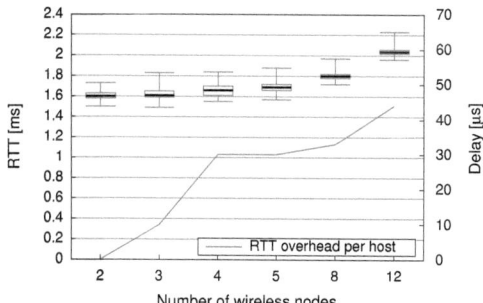

Figure 4.17: *WlanModel* scalability (distance = 300m, transmission interval = 1s, payload size = 56B).

4.8. EVALUATION

Multi-Hop Latency

For developments in wireless mesh networks, accurate results in multi-hop scenarios are crucial. The second scalability scenario, therefore, tests the multi-hop behaviour of VirtualMesh. It consists of several wireless nodes placed in a row. Static routing and interspaces of 500 m guarantee that packets are passed from node to its direct neighbour towards the destination. As Figure 4.18 shows, the results received using VirtualMesh fit the ones using pure simulation. An additional latency of the emulation can be observed for connections with more than six hops. It increases with the number of hops as each hop involves two additional transmissions between simulation model and the *iwconnect*. VirtualMesh can perfectly imitate the multi-hop behaviour of a wireless network. Therefore, it can be seen as a valid testing infrastructure. The remarkably higher standard deviation of the simulation results is caused by differences in the ARP implementation of the simulation and in reality. Once again, this shows that the emulation approach, with integration of the real protocol implementation, has significant advantages over pure simulation. To sum up, the received results attest VirtualMesh a good behaviour considering scalability. Connection with up to ten hops can be handled without any restriction when using the current emulation model and virtualisation.

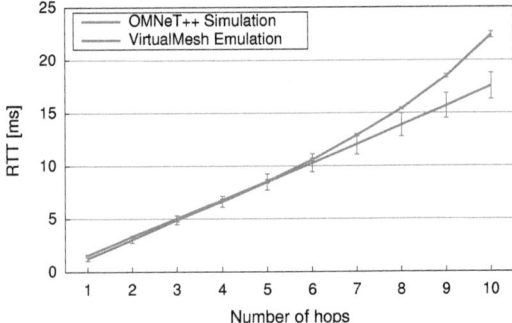

Figure 4.18: *WlanModel* multi-hop behaviour (distance = 500m, transmission interval = 1s, payload size = 56B).

Bandwidth

Another important metric for a network connection is its bandwidth. Throughput measurements have been performed with the tools *netperf* (TCP_STREAM) [102] and *nc* [82] in the emulation, and with extended versions the existing simulations

4.8. EVALUATION

EtherClientApp, TCPSessionApp and the UDPBasicApp in the pure simulation. TCP throughput is measured by the completion time of 200 MB bulk transfers. UDP traffic is measured by the UDPBasicApp sending packets of 1 MB size in short intervals to saturate the link. The total amount of transferred data after 300 s is then used to calculate the maximal UDP throughput.

Like for the RTT measurements, the impact of the infrastructure network and the traffic redirection is quantified between two end systems. The measured bandwidths with *netperf* for the 1 Gbps cross link are 904.56 Mbps to the native host and 578.64 Mbps to a virtualised host. The traffic redirection halves the values to 546.4 Mbps and 305.52 Mbps. Unfortunately, the traffic redirection with *iwconnect* does not scale with the 1 Gbps cross link. But, the provided bandwidth is still more than sufficient for experiments with an IEEE 802.11b wireless network with a maximal bandwidth of 11 Mbps (1.375 MB/s), or an IEEE 802.11a/g wireless network with a maximal bandwidth of 54 Mbps (6.750 MB/s).

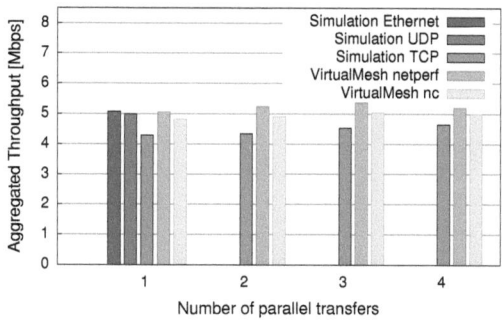

Figure 4.19: Aggregated throughput for parallel transfers using TCP and UDP.

After quantifying the impact of the infrastructure network and the traffic redirection, concurrent transmissions, and the multi-hop transmissions are measured. Figure 4.19 shows the measured aggregated average throughput in scenarios with 1-4 concurrent streams. The received results correspond to the net bandwidth that can be achieved in an IEEE 802.11b network [159] due to random transmission delays of MAC protocol CSMA/CA implementation and individual frame acknowledgements. There are no results for pure Ethernet transmission and UDP for concurrent streams as both are lacking any congestion avoidance mechanism, and would lead to full saturation with only one stream. The TCP implementation of OMNeT++ discards some valid ACK messages in wireless simulations, resulting in a reduced TCP throughput in the simulation. The problem has been confirmed by the OM-

4.9. CONCLUSIONS

NeT++ developers for wireless simulations, but it could not be fixed so far. This shows the benefit of VirtualMesh using the real TCP/IP stack of Linux. To sum up, the results in Figure 4.19 attest VirtualMesh realistic scalability in terms of parallel streams.

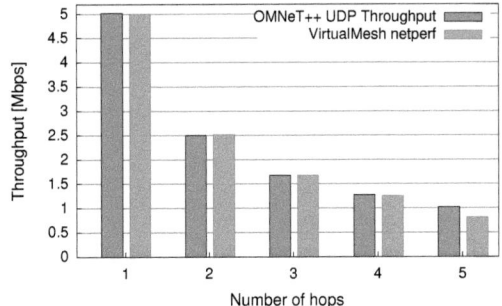

Figure 4.20: Multi-hop throughput results.

A final experiment measures the achievable bandwidth in multi-hop scenarios (see Figure 4.20). The nodes are placed in a chain topology, as in the RTT multi-hop measurements. VirtualMesh almost exactly matches the results of the simulation for up to four hops. It provides the expected behaviour. The end-to-end throughput decreases proportionally with the number of intermediate nodes. For more than four hops, VirtualMesh provides slightly lower bandwidth values than the simulation. To sum up, VirtualMesh matches the simulation results and provides correct values for maximal throughput, concurrent streams, and multi-hop communication.

4.9 Conclusions

In this chapter, VirtualMesh has been proposed as a new testing architecture that simplifies pre-deployment testing and development for protocols and architectures for WMNs and MANETs. After development and evaluation with network simulators, wireless mesh communication solutions require extensive pre-deployment testing of their target platform implementations. This is difficult to achieve in a real testbed, as irrepressible sources of interference exist. Furthermore, the variety of testing topologies is limited and mobility tests are impracticable. Therefore, we have designed VirtualMesh as a new testing architecture to be used before going to a real testbed. VirtualMesh is based on interception of wireless traffic at nodes

	Simulation	Testbeds	VirtualMesh
Transparency	-	+	+
Portability of protocols	-	+	+
Correctness of network behaviour	-	+	+
Scalability of tests	+	-	+
Repeatability of tests	+	-	+
Mobility tests	+	-	+
Development/equipment costs	+	-	+

Table 4.3: Qualitative analysis of simulation, testbeds, and VirtualMesh.

and redirection to a simulation model that provides more flexibility and a controllable environment. Table 4.3 summarises the benefits of VirtualMesh compared to simulation and evaluation in testbeds.

In comparison to other solutions (see Section 2.3), VirtualMesh provides a high integration of network emulation to real and virtualised hosts. The wireless drivers of the nodes are replaced by a virtual device that redirects traffic to an OMNeT++ simulation model instead of transmitting it over the air. This is fully transparent to the Linux network stack and the applications. The normal network stack and all applications can be used without any modification. Furthermore, even the standard configuration utilities can be used for wireless network configuration, as the virtual driver acts in the same way as in a standard wireless network driver under Linux. All configuration parameters may be set using common configuration tools. SliceTime [199, 201], a wireless network emulation extension for ns-3, has recently adopted our approach of a virtual wireless device driver.

VirtualMesh is used as an intermediate step before going to a real testbed and the final deployment. It offers the evaluation of the real protocol implementations on nodes running the final operating system and network stack in a controlled environment. Therefore, it can deliver results that match the network behaviour in a real testbed. VirtualMesh offers a scalability of tests comparable with pure simulation when using host virtualisation. It also provides good repeatability even though it cannot offer the repeatability of a network simulation, which delivers exactly the same results when started with the same parameter set. Node mobility in VirtualMesh is supported by the simulation model and can be defined either by different mobility models or by mobility traces. The high flexibility in setting up different test scenarios and host virtualisation drastically reduce the costs for testing the real implementations.

The focus of VirtualMesh is on the evaluation of wireless networks. It supports all common WMN scenarios, such as community networks and Internet sharing

4.9. CONCLUSIONS

with virtual nodes acting as gateways. Although VirtualMesh has primarily been designed to test WMNs, it is not limited to them. The concept can also be applied to other wireless networks, e.g., mobile ad-hoc networks (MANET).

A main advantage of VirtualMesh is the dynamic node management. Node registration and de-registration provides testing facilities even for management and software distribution frameworks that require rebooting of nodes, e.g., to update the operating system kernel or the communication software. During shutdown, a node just de-registers from the model and is not available until it registers again after being re-started.

We have evaluated the performance impact of VirtualMesh on delay and throughput. Our experiments have proven the full functionality of the VirtualMesh testing infrastructure. VirtualMesh introduces only negligible additional delays for traffic redirection and per real node inside a simulated path (0.35 ms per hop). The major part of the delay is caused by the infrastructure network - only a small fraction by the simulation overhead. Although VirtualMesh cannot accurately model effects that last for only few microseconds, such as difference in propagation delays depending on distance due to traffic redirection in user-space and the soft real-time scheduler, it is still a valid tool. Our experiments show that VirtualMesh is still able to realistically model real-world conditions in the milliseconds range. In contrast to simulations, it avoids problems of incomplete and inaccurate protocol implementations as a real Linux network stack can be used. The multi-hop behaviour of VirtualMesh accurately models the expected real world behaviour and matches the simulation. Our scalability tests have shown that VirtualMesh generally scales well. The achieved maximum bandwidth for all scenarios matches the simulation results. VirtualMesh models an IEEE 802.11b network accurately. The simulation model in VirtualMesh could be easily extended to provide support for IEEE 802.11a/g/h. As VirtualMesh does not introduce additional limitations, it is a perfect tool for protocol developers and practitioners to test developed software for WMNs prior to actual deployment.

Similar to other emulation frameworks, one problem of VirtualMesh is that the simulation may be too slow and cannot keep pace with the injected network traffic. Thus, the simulation model of VirtualMesh demands to be run on a powerful machine and to communicate over a distinct and high-performance management network. However, there may still be an overload situation. Therefore we plan to integrate the concept of synchronised network emulation [200] into VirtualMesh.

Even though VirtualMesh supports dynamic networks with nodes joining and leaving, as well as modifications of wireless parameters, some parameters such as the node positions and mobility pattern have to be pre-defined in the simulation model by using either existing mobility models or mobility traces. This limitation can be removed by propagating position updates through configuration messages to the simulation model in the future. Currently, the propagation of wireless parameters is

4.9. CONCLUSIONS

unidirectional from the nodes to the simulation server. Delivering return values from the simulation model is very interesting for future work as it could provide support for passive scanning of a wireless interface in the promiscuous mode (channel sniffing) and retrieval of SNR values. Another possible enhancement is the full support of highly dynamic multi-channel scenarios.

Currently, the simulation model supports nodes with virtual IEEE 802.11b network interfaces. In order to test other network technologies, they have to be added to the simulation model of VirtualMesh. The emulation model could then be extended to automatically select the correct simulation model for the used virtual wireless network device on the node including the corresponding individual parameter set. The client tools could be enhanced to support different wireless interfaces such as, e.g., IEEE 802.15.3, Bluetooth, or WiMAX. In [55, 30] we have expanded the VirtualMesh concept to the area of wireless sensor networks. A prototype is currently under development.

To employ VirtualMesh in the area of wireless sensor networks, the wireless model has to be extended. The wireless model provides almost unlimited possibilities for extensions. For example, the advanced MiXiM framework [202] with sophisticated MAC and physical models could replace the currently most frequently used INET models. Another future extension is the migration of the VirtualMesh client tools to the new Netlink-based wireless configuration interface of the Linux kernel. This would enable testing networks with the upcoming IEEE 802.11s standard.

VirtualMesh is released as open-source software under General Public License version 2 and available for download [178].

Part I introduced ADAM and Virtual as general frameworks and tools for supporting the life cycle of WMNs. In Part II (Chapters 5-7), we apply our developments to different WMN application scenarios, including a WMN for environmental monitoring, an ad-hoc WMN for video conferencing on construction sites, and a flying WMN for disaster recovery management. Moreover, we present our experiences and developed application-specific tools.

Part II

Application Specific Use Cases

Chapter 5

WMN for Environmental Monitoring

In this chapter, we describe our outdoor deployment of a WMN for environmental monitoring, funded by the Swiss Commission for Technology and Innovation (CTI)[169, 170, 171, 175]. Henceforth, we call our outdoor WMN deployment CTI-Mesh. It evaluated the utility and feasibility of WLAN-based WMNs in application scenarios, where remote sites need to be connected to a fixed broadband network. Examples for such scenarios are high-bandwidth multimedia sensor networks deployed in areas where fixed broadband networks have not yet been deployed or where it is considered too costly to deploy them. It has been tested whether the used hardware and software components are appropriate for the intended application scenarios. A deployment of a typical real world application as an outdoor testbed has been realised in the area of Neuchâtel Payerne, Switzerland.

Our contribution is the real deployment of a working 5 GHz WMN outdoor testbed using directional antennas with links up to 14 km in a rural area, whereas existing deployments often focus urban areas, e.g., MIT Roofnet, Berlin RoofNet, Heraklion Mesh. In contrast to urban networks, where nodes can be mounted on rooftops, the deployment in a farmland area requires that the nodes are solar-powered and mounted on masts, similar to QuRiNet network. Our deployment proved that the selected hardware and software components are suitable for the target scenario. Moreover, we share our valuable experiences in order to facilitate similar WMN setups in the future. As with any real-world deployment, many unexpected challenges arose prior to and during network setup and operation that demanded timely fixes and design decisions. Our documented experiences and best practises provide a good starting point for any future WMN outdoor deployment in rural and mountainous areas. In addition, we present evaluations of our deployed network, which was operational for about three months in 2009.

In CTI-Mesh, we proved the feasibility of a wireless mesh access network connecting remote locations to a fibre backbone network. CTI-Mesh gives control back to the network users. A company, researchers, or a community can deploy their own

5.1. INTRODUCTION

broadband access networks. Inexpensive broadband access is, therefore, not limited anymore to areas, where commercial operators see their profits. It can be brought to any remote area where it is required.

First, we introduce the CTI-Mesh project in Section 5.1 as well as the selected scenario for our outdoor feasibility test in Section 5.2. Afterwards, we present the used equipment in Section 5.3. Then, in Section 5.4, based on the regulations and equipment, we calculate important scenario parameters like the maximum permitted output power for the wireless network interface cards, minimum antenna/mast heights, and the expected received signal strengths. Section 5.5 presents the used operating software of the WMN. Afterwards, we proceed with the description of the planning and deployment in Section 5.6, which includes valuable experiences made during the planning and deployment. Section 5.7 shows our evaluation of the CTI-Mesh network and, finally, Section 5.8 concludes this chapter.

5.1 Introduction

Several application scenarios, such as environmental monitoring or meteorological data acquisition, require network connectivity in remote locations. This broadband network connectivity is not yet ubiquitously available; it might not be deployed in some areas due to commercial reasons, it does not deliver enough data bandwidth, or it is just too expensive. A cost-efficient network technology is therefore a necessity for the connection of remote locations and sensors to the broadband network.

The technology transfer project CTI-Mesh funded by the Swiss Commission for Technology and Innovation (CTI) and the participating industry partners evaluates the usefulness and feasibility of WLAN-based WMNs, where remote sites have to be connected to a fixed broadband network. The WMN should provide robust, reliable broadband network access guaranteeing a sufficient quality of service (QoS) for connecting high-bandwidth multimedia sensors for environmental monitoring. Besides the University of Bern, three industry partners, MeteoSwiss, SWITCH, and PCEngines, with different interests were involved in the CTI-Mesh project.

5.2 Application Scenario

CTI-Mesh connected a weather station located at Payerne to the fibre backbone with an access point at Neuchâtel. Due to the topography in the selected area (hills), forests and buildings, there is no direct line of sight between the network end points. A camera sensor had to be made accessible over a wireless mesh access network to the Internet by two paths utilising redundancy concepts in order to provide robustness and reliability (see Figures 5.1 and 5.2). The network consisted

5.2. APPLICATION SCENARIO

Figure 5.1: Map of Switzerland with the location of CTI-Mesh network.

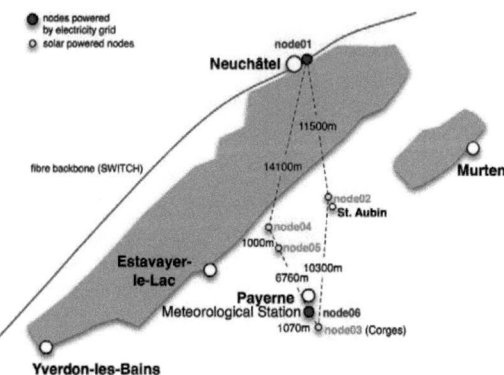

Figure 5.2: CTI-Mesh network deployed in the area Neuchâtel - Payerne, Switzerland.

5.3. EQUIPMENT

of six nodes, of which the four intermediate nodes were solar-powered (see Fig. 5.4 for an intermediate node). One end point of the wireless mesh access network, *node01*, was mounted on the rooftop of the University of Neuchâtel. It acted as gateway to the fibre backbone. The other end point, *node06*, operated as gateway to the sensor network with an IP capable camera. The four remaining nodes connected the two gateways and established two independent and redundant paths (see Figure 5.2).

(a) Node03 in Corges. (b) Node04 in Belmont.

Figure 5.3: Deployed nodes.

5.3 Equipment

In the following, we describe the equipment used for our outdoor deployment. It includes the mesh nodes, antennas, electrical power supply, mast, mounting material, and tools. Our equipment list is a good starting point; it might facilitate any future deployments of outdoor wireless mesh networks.

5.3.1 Mesh Nodes and Antennas

The PCEngines Alix.3D2 embedded board forms the core of our mesh nodes (see Section 2.2.1). Its two miniPCI slots hold two IEEE 802.11a/b/g/h cards. As

5.3. EQUIPMENT

Figure 5.4: Node05 deployed near Belmont.

Figure 5.5: Node06 on the platform roof of the MeteoSwiss building in Payerne.

5.3. EQUIPMENT

secondary storage, a 1 GB CompactFlash card is used. To keep accurate date and time in case of reboots or power outages, a small backup battery is added to the board to power the real-time clock (RTC). The node is packed in an aluminium weather sealed outdoor enclosure (see Figure 5.6). The enclosure fulfils the ingress protection standard IP67 [70], i.e., the enclosure is dust-tight and protected against the effects of temporary immersion in water.

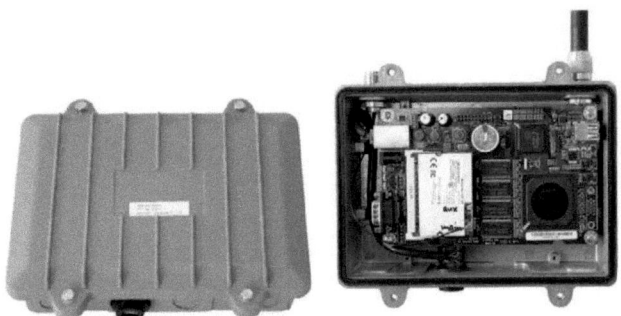

Figure 5.6: Water protected enclosure.

Two directional panel antennas (23 dBi gain, 9° beam width) are connected through 0.5 m low loss antenna cables (1.62 dB) and N-type pigtails to the wireless cards. The node's Ethernet interface is extended outside of the enclosure by a weather sealed Ethernet jack. A twisted pair cable then provides electric power and network connectivity to the node.

5.3.2 Power Supply for the Mesh Nodes

The mesh nodes in CTI-Mesh were powered either by the electricity grid or by solar panels. The two nodes mounted on the buildings of the University of Neuchâtel and MeteoSwiss (node01, node06) were connected via a lightning protector and a power over Ethernet (PoE) adapter to the standard electricity supply. The four intermediate nodes were supplied with electricity by solar power equipment. Besides a 80 W solar panel, the equipment consisted of an aluminium supply box, a solar charger, an acid battery (65 Ah, 12 V), and a passive Power Over Ethernet (PoE) adapter (see Figure 5.7).

A twisted pair cable to the electricity supply box connected the node on top of the antenna mast. This cable also provided network connectivity over Ethernet for on-site maintenance, which proved to be useful throughout the deployment and

5.3. EQUIPMENT

operation phase. In compliance with best practises, we mounted the solar panel vertically, which on one hand reduced the efficiency of the panel, but avoided other energy harvesting problems due to leaves, dust, rain, snow, and icing. The battery was dimensioned to support the self-sustaining node operation without being recharged by the solar panel for about ten days. During normal operation, the measured power consumption of a mesh node was approximately 3.3 W (271 mA, 12 V).

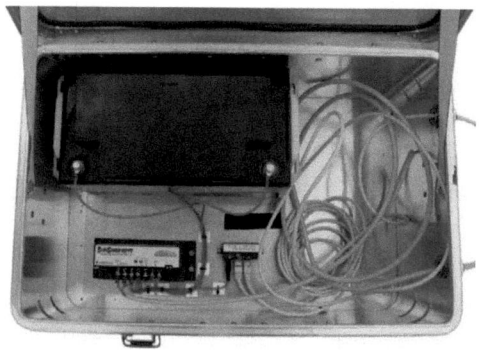

Figure 5.7: Power supply box with solar charger, lead acid battery, passive PoE adapter, yellow electric cable, and twisted pair cable.

5.3.3 Masts

Telescopic masts (sideways slotted aluminium tubes, max. height 9 m) with tripods were used to install the directional antennas and the mesh node in order to minimise disturbance and building activities. The mast type has been selected considering costs, transportability, project duration, and higher acceptability for the landowners providing the node sites for the installations. A mast tripod and a rope guying held the telescopic mast. We weighted the tripod with sand bags in order to get a basic stability of the mast. Iron stakes further fixed the tripod to the ground. The mast was guyed on two levels, each with three ropes. We selected a braided polyester guy rope with low stretch and easier handling rather than a steel guy wire. A first rope equipped with thimbles and wire clamps on both sides was connected with S hooks to the guying clamp on the mast and to the rope tightener. Then, a second rope was attached to the other side of the tightener and thereafter fixed to the ground by a wooden pile.

5.3. EQUIPMENT

5.3.4 Wall Mounting

The above described mounting support was used for all nodes except the node on the platform roof of the University of Neuchâtel. There, we mounted the antennas and the mesh node on a L-tube that has been anchored to the wall (see Fig. 5.8). Mounting of the antennas and nodes required several small parts like U-bolts, screws, and nuts.

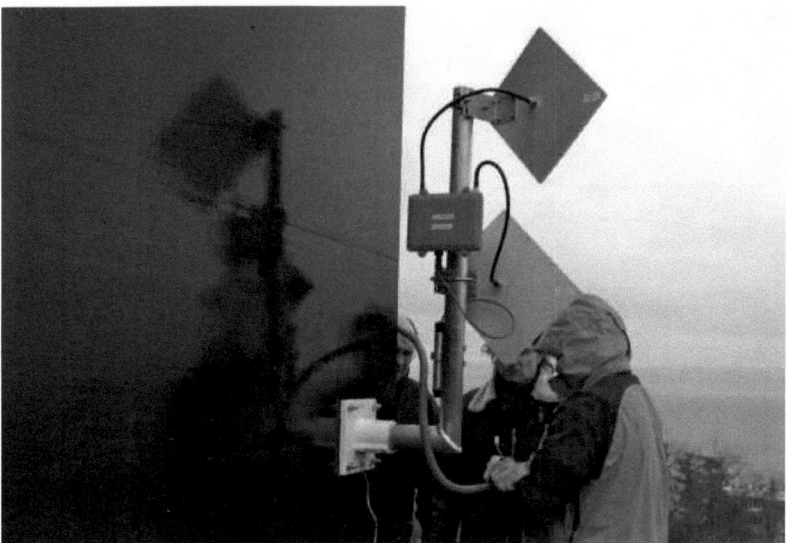

Figure 5.8: Assembling *node01* on the roof of the University of Neuchâtel.

5.3.5 Tools and Utilities

In order to assemble and mount the mesh nodes for the feasibility study, different tools were required (see Figure 5.9). The most important ones were a sledge hammer, slotted and Philips screw drivers, different wrenches, Allen keys, water pump pliers, a hammer, a knife, an angle measurement plate protractor, binoculars, a clinometer, an amplitude compass, a digital Volt/Ampere meter, a RJ45 crimp tool, a tester for twisted pair cables, a mast level, and two carpenter's levels. Moreover, a socket wrench with ratchet handle made life easier. A foldable ladder was useful as well. A sack barrow helped transporting the material and relieved the back. Finally, a folding chair made on-site configuration tasks more comfortable.

5.4. DEPLOYMENT PARAMETERS

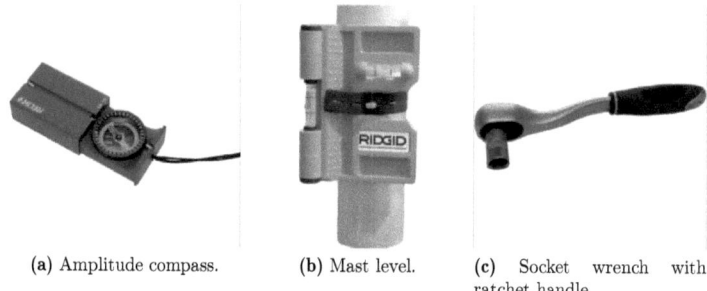

(a) Amplitude compass.   (b) Mast level.   (c) Socket wrench with ratchet handle.

Figure 5.9: Helpful special tools.

5.4 Deployment Parameters

During the planning phase of the project, we calculated the relevant deployment parameters for our setup, such as maximum transmit power, minimal antenna heights, and expected received signal power levels. These include the maximum permitted output power of the wireless network interface cards to comply with regulations, the minimal required antenna heights to guarantee good connectivity, and the expected received signal power levels to cross-check during the deployment.

Swiss regulations released by OFCOM limit the maximum transmission power to a value of 1000 mW EIRP when using TPC (see Section 2.7). EIRP [96] is defined as the emitted transmission power of a theoretical isotropic antenna to produce the same peak power density as in the direction of the maximum antenna gain. It is calculated by subtracting cable losses and adding the antenna gain to the output power (see Equation 5.1). The received power level at the receiver input (S_i) is shown in Equation 5.2. For our calculations we used the Free Space Loss (FSL) propagation model [75] as defined in Equation 5.3.

$$EIRP = P_{out} - C_t + G_t \tag{5.1}$$

$$S_i = P_{out} - C_t + G_t - FSL + G_r - C_r \tag{5.2}$$

whereas
$EIRP :=$ Equivalent Isotropically Radiated Power in dBi
$S_i :=$ Received power level at receiver input in dBm
$P_{out} :=$ Transmitted output power in dBm
$C_t :=$ Transmitter cable loss/attenuation in dB
$G_t :=$ Transmitting antenna gain in dBi
$G_r :=$ Receiving antenna gain in dBi

5.4. DEPLOYMENT PARAMETERS

$FSL :=$ Free Space Path Loss in dB
$C_r :=$ Receiver cable loss/attenuation in dB

$$FSL = 10\ log((\frac{4\pi}{c}df)^2) = 20\ log((\frac{4\pi}{c})df) \quad (5.3)$$

$$= 20\ log(d) + 20\ log(f) + 20\ log(\frac{4\pi}{c}) \quad (5.4)$$

$$= 20\ log(d) + 20\ log(f) - 147.55 \quad (5.5)$$

whereas
$FSL :=$ Free Space Path Loss in dB
$f :=$ Frequency in Hz
$c :=$ Speed of light in a vacuum 300'000'000 m/s
$d :=$ Distance between transmitter and receiver in m

It is required that at least 60% of the first Fresnel zone are free of any obstacles in order to use the FSL model for calculation of the attenuation. Otherwise, additional attenuation has to be added. Equation 5.6 calculates the radius of the zone that has to be free around the line of sight. The earth's curvature is a further obstruction of the Fresnel zone. Hence, the minimum antenna height has to consider it as well. Equation 5.7 defines the additional antenna height EC_m due to the earth curvature [191]. It also considers the effect of atmospheric refraction, which causes ray distraction at microwave frequencies. In practice, the reception of the microwave signal is possibly somewhat beyond the optical horizon. The minimum antenna height H_{min} is then defined in Equation 5.8. For our calculations in Table 5.1, we used the values $EIRP = 30dBm$, $f = 5.5GHz$, $C_r = 1.62dB$, and $C_t = 1.62dB$.

$$FZ_{r(m)} = 0.6 \times \frac{1}{2}\sqrt{\frac{d \times c}{f}} \quad (5.6)$$

$$EC_m = \frac{d_1 \times d_2}{12.8 \times k} \quad (5.7)$$

$$H_{min} = EC_m + FZ_{r(m)} \quad (5.8)$$

whereas
$FZ_{r(m)} :=$ Radius for 60% of the first Fresnel zone
$EC_m :=$ Additional antenna height due to earth curvature
$d_1, d_2 :=$ Distances point ↔sender/receiver in km.
$k := \frac{4}{3} \times$ earth radius (6'371 km)

As all our node sites were located on top of hills, our telescopic masts with a height of 9 m were sufficient to guarantee no obstacles in the first Fresnel zone. Keeping the antenna heights below 10 m, further avoids the necessity to request building permission from the local authorities.

5.5. SOFTWARE

Table 5.1: Links using 1000mW EIRP.

$Node_{xx}$	d_m	$FZ_{r[m]}$	$H_{min[m]}$	$FSL_{[dB]}$	$S_{i[dBm]}$	$P_{out[mW]}$
01 ⇔ 02	11500	7.513	9.463	128.47	-77.09	7.277
02 ⇔ 03	10300	7.110	8.668	127.51	-76.13	7.277
03 ⇔ 06	1070	2.291	2.308	107.85	-56.46	7.277
06 ⇔ 05	6760	5.760	6.431	123.86	-72.47	7.277
05 ⇔ 04	1000	2.215	2.223	105.26	-53.87	7.277
04 ⇔ 01	14100	8.319	11.239	130.24	-78.86	7.277

5.5 Software

For the CTI-Mesh deployment, we used ADAM (see Chapter 3), our own Linux distribution, as operating system for the mesh nodes. The configuration of the wireless network interface of ADAM was extended to be compliant with the Swiss regulations (see Section 2.7) in terms of maximal transmit power, TPC and DFS. ADAM facilitated the network configuration of the entire network, as all relevant parameters are hold in only one single configuration file per node. Using ADAM, we could further safely update the software and configuration of the WMN nodes without requiring a co-located management network. Failed software update were automatically recovered and did not lead to broken nodes, which would have required time-consuming disassembling of the masts.

As wireless driver, we used a patched version of MadWifi 0.9.4 [186]. Other parts of the communication software are the Linux IPv4/IPv6 dual stack and a routing daemon. ADAM already fully supports IPv4 and IPv6. The routes inside the wireless mesh network are automatically established by the olsrd routing daemon [187], an open source implementation of the OLSR with ETX routing metric (see Section 2.1.1).

A concurrent IPv4 and IPv6 configuration has been selected for the CTI-Mesh network. Public IPv4 and IPv6 addresses have been assigned to every wireless interface in the network. In addition, the gateway node (*node01*) in Neuchâtel and the mesh node (*node06*) in Payerne have public IP addresses assigned to their Ethernet interface enabling access to either the fibre backbone or the IP webcam. The network could also have been setup with network address translation for the IPv4 addresses at the gateway node. However, due to easier accessibility, all nodes used public IP addresses. Every intermediate mesh node ran a DHCP server providing private addresses on its Ethernet interface for on-site maintenance.

5.6 Planning, Predeployment, and Deployment

A field test requires several steps in planning and predeployment. We recommend the following actions as our best practise: time planning, selection of testing area, finding appropriate locations for intermediate nodes, reconnaissance of node sites, agreements with landowners, determining and ordering appropriate equipment and tools, preparation of equipment, setup of software and configuration, pre-deployment tests, and final deployment.

A complex project with several external dependencies requires extensive time planning and scheduling. One has to consider the availability of means of transportation, equipment, and external parties, such as public administration and landowners. Further restrictions may be caused by site accessibility and prevailing weather conditions.

Besides a time schedule, a testing area and the elevated node sites providing line-of-sight connections are required. Accurate electronic maps help to determine candidate locations for the deployment. As there are always differences between maps and reality, a next step is to go on-site (reconnaissance) and verify whether the sites are actually useable. Then, the landowners have to be contacted in order to get permission for using their property for the tests. For getting the agreements, we had the best experiences when talking face-to-face.

Another activity is checking and preparing the equipment. Once the ordered equipment has been delivered, completeness and functionality should be checked. It is then advisable to prepare the material before going in the field, e.g., assembly of nodes and antenna, preparing guying ropes by cutting them and adding thimbles and wire clamps.

The next step should be a predeployment test. All equipment is assembled completely and set up outdoors. This helps in identifying defective and missing parts. Moreover, first stability tests of hardware and software can be performed.

After the predeployment tests, one can proceed to the final deployment. Of course, there are problems that arise after the planning and predeployment phase. The next section gives an overview of different challenges that occurred during our entire deployment.

During the deployment, we had to find practical solutions to several problems and challenges. We classify the challenges into the following five categories:

- Software problems

- Mechanical challenges, missing or defective material

- Technical communication problems

- Natural environment

5.6. PLANNING, PREDEPLOYMENT, AND DEPLOYMENT

- Administrative challenges

Software Problems

Some software problems occurred during the project. First, the outdoor use of IEEE 802.11h (TPC and DFS) in combination with ad-hoc mode is not commonly used in the community and, therefore, not the highest priority for the developers of the wireless driver. Thus, the wireless driver provided poor support for these configuration settings, i.e., it was not extensively tested, and the implementation had several errors. By applying several patches from the OpenWrt project [14], we significantly improved the system's stability and operation. Second, the routing daemon stopped working occasionally. Monitoring the routing daemon and restarting if necessary solved this problem.

Mechanical Challenges

The mechanical challenges included correct antenna alignment at setup, sinking in of tripods, torsion of mast elements by fixed guying clamps, and defective material. The correct alignment of the antennas is crucial as directional antennas were used. After having calculated the angles and elevations by using maps, there were four mechanical problems for correct alignment.

First, the two antennas had to be fixed to the top mast element with the correct intermediate angle. We adjusted the pre-calculated angle using a precision mechanic universal Bevel protractor.

The second problem was keeping the exact direction of one antenna aligned to a reference system on the bottom element of the telescopic mast, which is then at eye level. Any attempt to lift the mast elements in vertical position resulted in torsion of the top element compared to the bottom element. Therefore, we assembled the mast completely in horizontal position and then erected it in one piece (see Figure 5.10). In order to transcribe the antenna direction to the reference plate, we used two carpenter's levels when the mast was in horizontal position. One carpenter's level was positioned on one of the antenna and balanced. The reference plate was then aligned and balanced with the other one. Using an amplitude compass on the reference plate, the antenna could then be aligned correctly. Since preliminary tests [175] revealed that visual alignments of the antenna failed, an amplitude compass and an inclinometer have been used for correct alignment. Afterwards, we fine-tuned the alignment with the help of the received signal strength. Although the alignment with the amplitude compass generally worked well when being in fields, there was magnetic interference from generators on the platform roof of the University of Neuchâtel, which required us to make several attempts for the correct alignment of the antennas of *node01*.

5.6. PLANNING, PREDEPLOYMENT, AND DEPLOYMENT

Figure 5.10: Complete assembly of telescopic mast in horizontal position before final setup.

The third mechanical challenge was the sinking in of the tripod into the soft and rain-sodden soil after heavy rain falls. The results were lopsided masts. Thus, we stabilised the ground with concrete paving slabs as shown in Figure 5.11). Another option is using aluminium tripod mount plates, which can also offer the possibility to easily equalise the inclination of the tripod.

The fourth mechanical challenge was an unexpected torsion of some mast elements, which occurred over time and resulted in wireless connection losses of the directional antennas. The cause of the torsion was the fixed mounted type of guying clamps used. On all node sites, not all of the guying ropes could be fixed with intermediate angles of 120°. Therefore, the ropes' tensions produce a torsional force, which then turned the mast element. Now movable guying clamps (fibre-enforced plastic discs) as shown in Figure 5.12 solved the problem by decoupling the mast elements and the guying.

In pre-deployment tests, the complete equipment was set up. The tests showed the necessity of two guying levels to avoid oscillations of the mast top holding the antennas. Moreover, they helped us to identify missing or defective material before the final deployment, minimising the consequences such as unnecessary on-site corrective actions and delays.

5.6. PLANNING, PREDEPLOYMENT, AND DEPLOYMENT

Figure 5.11: Concrete paving slab to prevent sinking in of the tripod, sand bag and iron stake to stabilise mast.

Technical Communication Problems

During the network setup two communication problems appeared. First, we discovered unexpected packet loss on the wired link between the border router and the gateway node *node01*. The dedicated twisted pair cable (100 m) in combination with the data link lightning protector produced high attenuation and collisions. Reducing the cable length to 50 m by taking advantage of the existing building wiring eliminated the problem and resulted in the expected 0 % packet loss on the wired link. Second, the different wireless links interfered with each other as they communicated on the same channel. The interference was reduced by alternating use of three channel sets and exploiting the two available antenna polarisations (horizontal and vertical).

Natural Environment

The natural environment had several influences on our feasibility study. Besides described problems due to the rain-sodden ground, fog, storms, and animals had an impact on the network. The solar panels used should have normally produced

5.6. PLANNING, PREDEPLOYMENT, AND DEPLOYMENT

Figure 5.12: Primarily used fixedly mounted threepart guying clamp and its replacement part, a movable guying clamp to prevent torsion of mast.

enough energy to charge the batteries and power the mesh nodes twenty-four-seven throughout the year, independent of weather conditions. Nevertheless, we observed two nodes that completely drained their batteries and thus stopped working for approximately one week in November 2009. The other two solar-powered nodes had completely charged batteries in the same period during daytime. In fact, bad weather conditions, including locally dense fog over several weeks, prevented the solar panels from producing enough energy to charge the batteries. Once the solar panel delivered again enough electric power, following the bad weather period, the nodes restarted normal operation without any intervention by an operator.

Furthermore, parts of our equipment were severely damaged during storms. First, lightning destroyed the web cam on the roof of the MeteoSwiss building during a thunderstorm. The mesh node was not affected due to the data line lightning protector. Second, a wind storm broke one of the masts as one guying rope had become loose (see Figure 5.13). As no further mast was buckled, even during heavier

5.6. PLANNING, PREDEPLOYMENT, AND DEPLOYMENT

Figure 5.13: Broken mast due to strong winds and loose guying (*node02*).

windstorms, we are convinced that the selected mast material is sufficient as long as the guying is correctly applied. Birds of prey (common buzzards) used our masts and antennas as raised hides/perches. Since they also sat on the antenna cables, they loosened the connector on the antenna. Tightening and gluing the connector reduced the effect. We did not succeed in keeping the birds away from the masts. Other animals taking profit of our installations, such as spiders, ants, beetles, and mice did not influence the network performance.

Administrative Challenges

The last category is formed by administrative challenges. First, we required the agreements for hosting a node. After the time-consuming determination of appropriate node sites and their landlords, convincing the landlord to give an agreement was demanding. Face-to-face communication and showing the equipment were the key elements for success. Second, determination of the suppliers for all the required equipment and tools was difficult and keeping track of all the parts and pieces is a

5.7 Evaluation

The aim of the deployment activities was to connect sensing equipment over a WMN to the wired/fibre backbone. As a show case application, an IP camera was connected and accessible from the Internet during the deployment (see Figure 5.14).

Figure 5.14: Screenshot of IP camera streaming over WMN.

In [175], we presented some preliminary measurements. During these measurements, strong winds caused periodic movements of the antenna tops which resulted in high packet losses. In the final deployment, guying the antenna to the ground with ropes eliminated this effect.

For all measurements, the CTI-Mesh network used a fixed data rate of 6 Mbps for the IEEE 802.11h interfaces. Setting higher data rates is possible, but the longest links stretching over 10 km may then become unavailable.

In order to give an impression of the achievable bandwidth over the deployed network, we performed TCP bandwidth measurements using the tool iperf [78]. These results are shown in Figure 5.15 and 5.16. The measurements were started in

5.7. EVALUATION

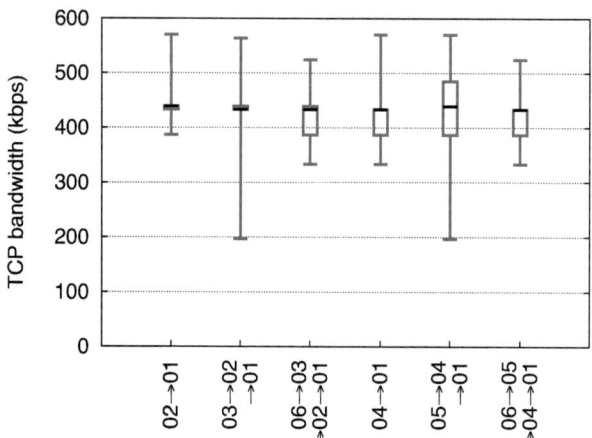

Figure 5.15: TCP bandwidth for the connections to *node01*.

sequence and lasted for 10 min. Data values were produced for periods of 10 s. In the graphs, the data is represented by its median value, the 25% percentile and the 75% percentile (box), and the minimum and maximum values (whiskers).

First measurements were run from the nodes towards the gateway (*node01*) (see Figure 5.15). The results are similar for all nodes, with a median value of 439 kbps. Due to the alternating use of the two available antenna polarisations (horizontal and vertical) and of three channel sets, there is almost no intraflow interference along the multi-hop path (02 → 01, 03 → 02 → 01, 06 → 03 → 02 → 01, 04 → 01, 05 → 04 → 01, 06 → 05 → 04 → 01). The bottleneck for the TCP transmissions was the link with the lowest bandwidth.

Figure 5.16 presents the second measurements, performed between direct neighbours. It shows that the overall bandwidth is mainly limited by the long distance links above 6 km. The capacity of the 1 km link between *node04* and *node05* reaches about 55% of the set data rate (6 Mbps), which lies slightly below the commonly reported throughput values. In fact, this link could not be positioned ideally. A bordering forest located in the middle of the link covered more than the 50% of the first Fresnel zone. The low value for the 1 km link between *node06* and *node03* may be explained by the fact that setting the correct elevation angle (3° due to the difference in altitude) for the antennas, was very difficult with our equipment. Moreover, the link is aligned directly with the city centre of Payerne. By a channel scan, we identified several neighbouring concurrent wireless networks that our

5.7. EVALUATION

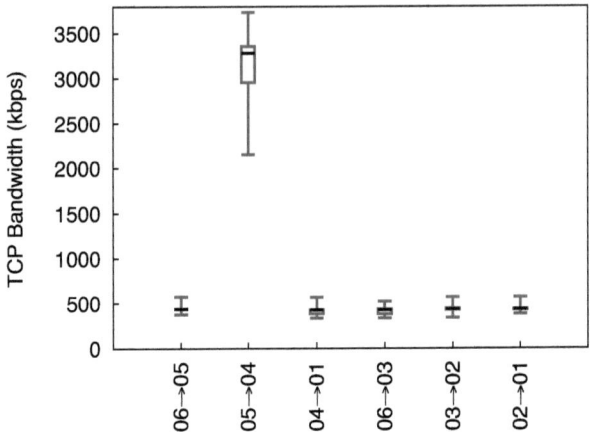

Figure 5.16: TCP bandwidth for each link.

directional antenna can receive and, therefore, resulted in interference.

In order to monitor the network's availability and the link/route quality, we logged the routes to *node06* with the corresponding routing metric ETX (Expected Transmission Count) cost values at *node01* every 10 min. This was performed using standard functionality of the olsrd routing daemon. ETX defines the number of transmissions that are required to successfully transmit a packet. In Figure 5.17, the weekly ETX values are depicted and show that most values are near to the optimum of 3.0 for the three hop path (*node01*↔*node06*). The whiskers show the minimum and maximum values. ETX values above 9.0 usually occurred when the connection was lost or after the connection became available again.

Figure 5.18 provides an overview of the general route availability towards node06 and the IP camera for 81 days. It shows the percentage of the day during which at least one valid route is available. Several events had an impact on the route availability, e.g., wind breaking the mast of *node02* on day 45, which was replaced nine days later. Moreover, stability problems of the wireless driver led to non-functioning wireless devices. The effect could be minimised by automatic service restarts and reboots after day 44. The drawback of some unnecessary restarts is that the maximal achievable route availability was reduced to about 99%. In many situations, this network performance may be sufficient, as most sensor data can be aggregated and then transmitted with some delay. Moreover, redundant paths can be used to cope with short link outages. By periodic ICMP ECHO measurements,

5.7. EVALUATION

we further measured the average delay and the corresponding packet loss on the path between *node01* and *node06*. After fixing the software issue and replacing the mast of *node02* (day 54), the measured average round trip time (RTT) was improved to 11.6 ms and the average packet loss to 7.18%.

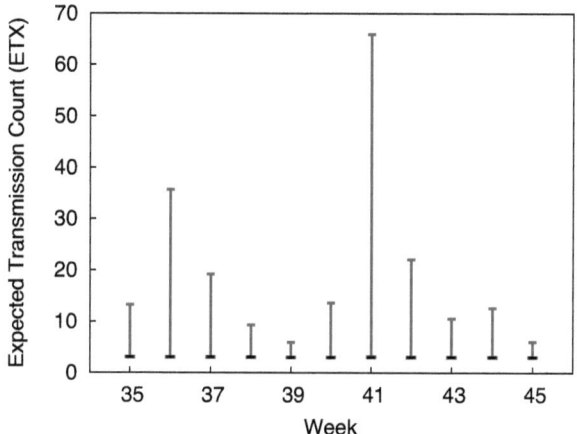

Figure 5.17: ETX values for best route from *node01* to *node06*.

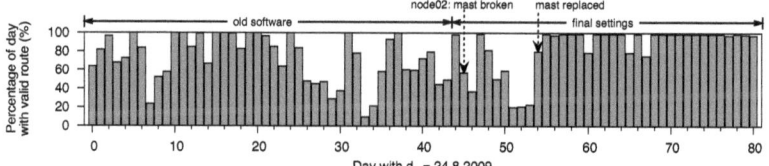

Figure 5.18: Route availablity to *node06*/IP camera at *node01*.

In order to verify our deployment, we logged the signal strength values at each node (see Figure 5.19). The resulting median values are symmetric for both directions of the same link and correspond to the calculated signal strengths $S_{i(dBm)}$ in Table 5.1. The variance of the results is partly due to TPC adjusting the transmission power.

Despite using alternating antenna polarisations, high quality cabling, orthogonal channels and channel separation, the network performance may still suffer from

5.8. CONCLUSIONS

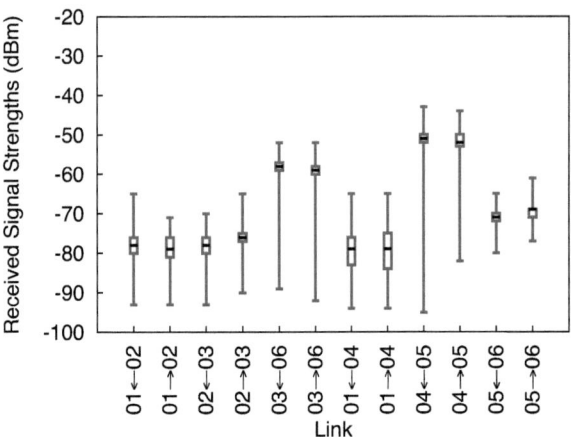

Figure 5.19: Received signal strengths for all six links.

adjacent channel interference and, if using multi-radio systems, board crosstalk and radiation leakage [7, 49, 131]. Although not observed in our setup, increased separation of the antennas and additional shielding is possible and recommended.

5.8 Conclusions

In this chapter, the outdoor deployment of a WMN for environmental monitoring (CTI-Mesh) has been discussed. The CTI-Mesh access network interconnected remote sensors, including a live cam, to a fibre based network backbone over a distance of more than 20 km.

We presented our deployment experiences for this solar powered wireless access mesh network for meteorological data acquisition. These experiences and the established deployment process provide a valuable starting point for any future WMN outdoor deployments. They help in being aware of common problems and pitfalls, directly avoiding them, and finally saving a lot of time. Our experiences show that one of the most important steps in the deployment process is predeployment testing. Extensive predeployment tests are crucial to avoid unnecessary delays or project failure. Therefore, we strongly advise to perform them carefully. Besides testing the communication software, it is advisable to set up the complete nodes including masts and solar equipment before on-site deployment. This enables iden-

5.8. CONCLUSIONS

tification of missing or defective equipment and tools before going into the field. Moreover, replacement parts should always be kept available. Otherwise, setup and repairs may be delayed by additional on-site operations or even by long delivery times for spare parts. The network deployment including all preparation steps took about three months.

We proved that a wireless access network is feasible and that the utilised equipment is appropriate for such deployments in the selected region. The double-guyed aluminium mast with the tripod placed concrete paving slabs resisted strong winds and thunderstorms. The electronic equipment was properly shielded against various weather conditions including rain and snowfalls. The utilised equipment is flexible and adaptable and, therefore, provides an appropriate solution for various terrains.

We further demonstrated that the network can be operated completely self-sustained, i.e., the network nodes can be powered by solar equipment consisting of solar panel, charger, and battery. This requires an appropriate dimensioning of the solar panel and the battery, which takes into account local weather effects, such as locally dense fog.

Concerning the feasibility and usability of access network, our evaluations of the CTI-Mesh network showed that our setup can provide a requested robust network service for transmitting weather data (430 kbps over 20 km). The network performance is sufficient for the use case. The network stability might be further improved, e.g., by replacing or extending the OLSR routing daemon to avoid route fluctuations and migration of the used MadWifi wireless driver to its successor driver, the completely open source wireless driver ath5k. Moreover, integrating a hardware watchdog that could recover a node from undefined states could enhance self-healing mechanisms.

To sum up, the CTI-Mesh deployment provides tested low cost equipment and software for an outdoor deployment of wireless mesh access network. It further delivers empirical data for the configuration and performance of an outdoor WMN. In addition, the established deployment process defining best practices and documented challenges with their solutions provide an excellent starting point for any similar outdoor network deployment.

CTI-Mesh employed ADAM (see Chapter 3) for network management. ADAM provided an embedded operating system specially tailored for the WMN nodes, offering all necessary functionality for the CTI-Mesh deployment. It simplified the configuration of the entire network, as all relevant parameters can be set in only one file per node. ADAM guaranteed the distribution of configuration and software updates even in situation where certain nodes were temporarily off-line, e.g., due to empty batteries. ADAM's safe software update procedure for the used ALIX nodes ensured that the software images could be safely exchanged. There was no risk of a broken node, which would have required a time-consuming disassembly of the

5.8. CONCLUSIONS

mast. Self-healing mechanisms increased the robustness of the deployed network. Using the ADAM management framework, CTI-Mesh did not require a co-located management network like other deployments, e.g., Heraklion Mesh.

Chapter 6

Deployment Support for an Ad-Hoc WMN

After the manual deployment of an outdoor WMN for environmental monitoring, we present two support frameworks for the deployment of ad-hoc WMNs in the following two chapters. These frameworks either guide the user in the deployment process or even offer a fully automatic deployment of a WMN (see Chapter 7). In this chapter, the first framework, called On-site Video System (OViS) [140, 182, 183], is presented. It offers semi-automatic deployment of a temporary network as a communication infrastructure for different support applications, e.g., an audio/video conferencing system. It guides an inexperienced user, by means of an electronic wizard, how to correctly deploy a temporary WMN for the purpose of video conference system.

OViS was developed motivated by a problem occurring at an electric installations company, which wanted to reduce the number of costly visits of its electrical engineers to construction sites. By using an audio/video conferencing system, engineers would be able to support multiple construction sites per day and further reduce time wasted through unnecessary travel. The audio/video conferencing system, i.e., video conference, requires network connectivity at the problematic sites. Unfortunately, there is usually no coverage by an already installed wired or by a cellular data network in the basements of buildings, where most electric installation problems occur. However, larger construction sites have an own Internet connection in one of the construction containers. The challenge is then how to extend this connection to the basements without introducing safety risks, such as tripwires.

OViS provides this connectivity by a temporary battery-powered WMN, whose deployment can be quickly performed even by inexperienced users. The user is guided by a wizard-like deployment application during the deployment process. By following instructions of this deployment application, the user sets up the network step-by-step. He/she is instructed to deploy the mesh nodes at reasonable distances based on received signal strength indicator (RSSI) measurements. OViS handles

6.1. MOTIVATION

the configuration of the WMN and completely hides this from the user. The OViS deployment wizard exists in versions for different systems and operating systems, including versions for Android and iOS smart phones.

Section 6.1 presents the motivation for OViS, which provides a semi-automated deployment of a temporary WMN for video conference. Following, we discuss the architecture and concepts in Section 6.2, the implementation on the mesh nodes in Section 6.3, and the implemented deployment applications (wizards) in Section 6.4. Section 6.5 provides an evaluation of our OViS temporary network and of the deployment process. Section 6.6 concludes this chapter.

6.1 Motivation

Nowadays, information and communication technology (ICT) has already brought significant cost savings to several industries, including the building construction industry. During the construction of a building, modifications may require costly on-site visits of engineers to adapt plans to the new circumstances. ICT solutions, such as video conferencing systems, may reduce the number of on-site visits significantly. Video conferencing enables the engineers to remain in the office and yet to comprehend problems and particularities of complex new situations, adapt their planning, and then instruct the workers on-site. Unfortunately, in-building communication networks, as well as electric installations, are set up very late in the building construction process. In addition, communication over cellular mobile networks (GSM/UMTS) is often not possible inside buildings, especially in basements.

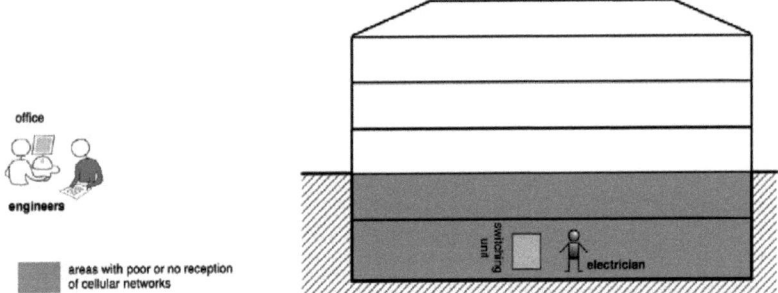

Figure 6.1: Motivation for OViS: An electrician requires instructions to solve an issue at the switching unit in the basement of a building. Unfortunately, there is no reception of cellular networks in the basement.

As deviations and adaptations to the plan are quite common during building

6.1. MOTIVATION

construction, an electric installations company's engineers often have to support their electricians in adapting the planning on-site. The costly engineers then spend a lot of their working time by travelling from the office to the construction site and back. The obvious solution of using a phone often fails due to the following two reasons. First, the situation may be too complicated to be explained on the phone. A picture or a video could illustrate a complex situation more easily. Second, the complicated electric installations including the switching units are usually situated in the basement out of reach of any cellular network (GSM/UMTS) (see Figure 6.1). Therefore, the electrician has to go outside the building in order to make a phone call to the main office or to send a picture or video. If more data is required, he/she has to return to the basement to check the circumstances or to make another photo. Afterwards, he/she has to leave the building again to communicate with the engineer. An audio/video conferencing system that enables communication whilst being next to the problematic installation would be beneficial and would give the company a competitive advantage.

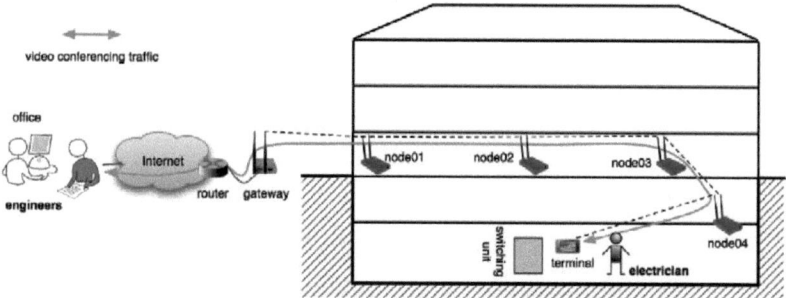

Figure 6.2: OViS: A temporary WMN provides Internet connectivity in the basement and, therefore, enables video-conferencing to discuss problems comfortably and efficiently.

A temporarily disposable communication infrastructure, which enables on-site audio/video conferencing functionality, would therefore constitute a much-appreciated benefit. It should be simple, straightforward, and safe in its deployment. As on-the-fly cable installations are safety risks on a construction site, we investigated the usability and applicability of a WMN with battery-driven nodes as a temporary communication infrastructure, as shown in Figure 6.2, and developed our on-site video-conferencing system (OViS). The application of WMNs for temporary venues and spontaneous networking has also been suggested in [4]. In order to provide "as easy as winking" deployment, OViS includes self-configuration and self-awareness mechanisms and guides the inexperienced user through the deployment by a wizard application.

6.2 OViS Concepts and Architecture

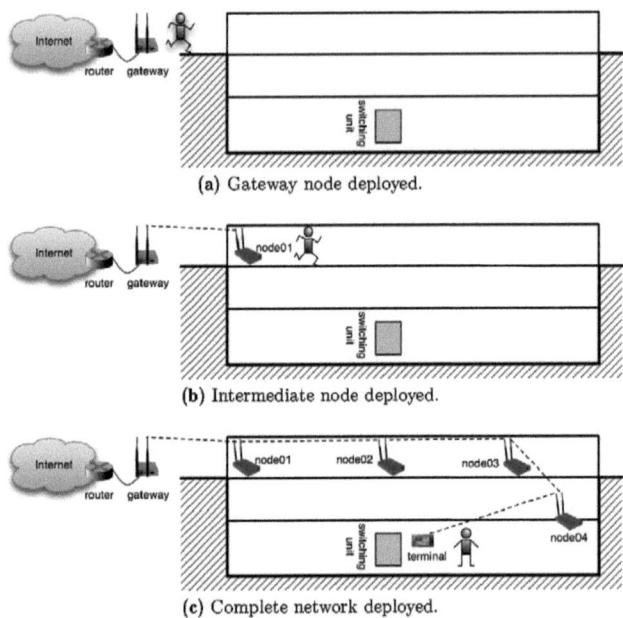

(a) Gateway node deployed.

(b) Intermediate node deployed.

(c) Complete network deployed.

Figure 6.3: Stepwise deployment of the temporary network for OViS.

Figure 6.3 illustrates the general deployment process of the temporary WMN in OViS. A user takes the bag containing the components of OViS, namely battery-powered mesh nodes, a client device (e.g., smart phone or tablet PC), and an Ethernet cable. He/she first deploys the gateway node near to the local router in the construction container (see Figure 6.3a). Then he/she takes the next node out of the bag, switches it on, and walks into the direction of the problematic switching unit. At a reasonable distance, the first intermediate node is deployed (see Figure 6.3b) and the user proceeds with the deployment of nodes until he/she reaches the switching unit (see Figure 6.3c). After the successful network deployment, the user is in front of the switching unit. He/she can start a video conference to discuss the problem and to clarify the next steps with an expert/engineer back in the office. The audio/video conferencing system helps in finding a solution more efficiently, as the engineer can grasp the problem a lot easier with live video/pictures and can

6.2. OVIS CONCEPTS AND ARCHITECTURE

instruct the electrician to check/show some parts of the system, if necessary. The engineer may even consult additional documentation in the office and give immediate feedback.

6.2.1 Requirements

In order to be usable as a communication system for audio/video conferencing, OViS has to fulfil the following requirements:

- Easily deployable by an inexperienced user
- Quick setup time
- Network bandwidth supporting video-conferencing systems (throughput $\geq$ 1 Mbps)
- Self-sustaining operation for hours
- Independent of localisation mechanisms

Following the first requirement, we cannot assume any prior knowledge about networking, especially wireless multi-hop networks, from the user. He/she is not aware how to deploy a wireless multi-hop network correctly, e.g., the ideal distance between the nodes or the configuration steps for a network interface. Nevertheless, he/she should be able to deploy the network correctly within a reasonable time (less than 10 minutes) at any place, where connectivity is necessary. This deployment process has to be flexible in order to work at different locations regardless of obstacles and interferences. Finally, the deployed network has to provide enough bandwidth for supporting a video conference and work for hours without additional user interaction, i.e., the nodes have to run battery-powered for a couple of hours. Since no absolute positioning information is available indoors or only with costly equipment, OViS has to support the deployment process without it.

Our solution is a deployment wizard application (OViS client) that guides the user throughout the deployment process, and a completely self-configuring WMN. The OViS client instructs the user to perform all necessary steps, such as connecting a node to the local router by an Ethernet twisted pair cable or switch a node on. It provides easily comprehensible instructions for the placement of the nodes, such as "move further away from the previous node". For delivering these instructions to the user, the OViS client interprets the measured RSSI values between the current node and the previously deployed one. In order to provide acceptable network performance and to reduce mutual interference between neighbouring links, OViS employs multi-channel communication.

6.2. OVIS CONCEPTS AND ARCHITECTURE

6.2.2 Network Setup

A user principally deploys an OViS network in a chain topology between the local router and the location of the problem just before he/she wants to communicate. Nevertheless, there are still several options for the network setup of OViS. In the following three design alternatives are discussed. In all design alternatives, we assume that the local router provides IP addresses using the dynamic host configuration protocol (DHCP) [68].

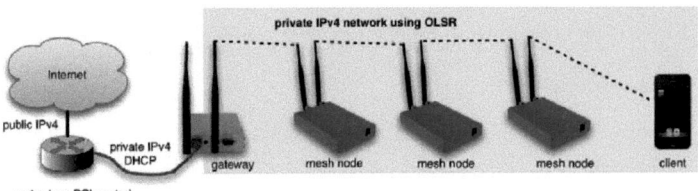

Figure 6.4: OViS network topology: Full OLSR (IPv4).

The OViS network could be setup by deploying an ad-hoc routing protocol on all OViS network participants, i.e., mesh nodes and the client (see Figure 6.4). In our case, OLSR is used as routing protocol. The full OLSR topology can either be IPv4 or/and IPv6. However, it would require OLSR to be installed on the OViS client, representing a major drawback of the approach. Although, OLSR implementations are existing and even available for mobile platforms, such as Android and iOS, they cannot be installed without voiding the guarantee of the smart phones by modifications. Therefore, this approach limits possible platforms for the OViS client and is completely discarded.

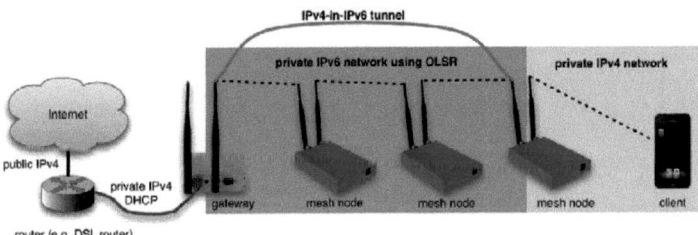

Figure 6.5: OViS network topology: OLSR (IPv6) with IPv4-in-IPv6 tunnel.

The second approach, shown in Figure 6.5, avoids this limitation by introducing an additional network just for the last node and the OViS client. It is based

6.2. OViS CONCEPTS AND ARCHITECTURE

on an IPv6-only mesh network with OLSR. In order to stay compatible with IPv4 Internet and applications, an IPv4 network for the OViS client is connected to the IPv4 network of the local router by an IPv4-in-IPv6 tunnel between the last node and the gateway node. As soon as IPv6 is widely deployed, the whole topology could be easily switched to IPv6 by removing tunnelling and IPv4 network parts. This would simplify the set up of video conference, as all network devices are properly addressable and directly reachable. Obviously, the approach suffers from the additional complexity for the tunnelling as long as IPv4 compatibility is required. Tunnelling includes additional headers to packets. In consequence, the maximum transmission unit (MTU) on the OViS client has to be manually decreased in order to avoid additional packet fragmentation, which results in a significantly lowered network performance.

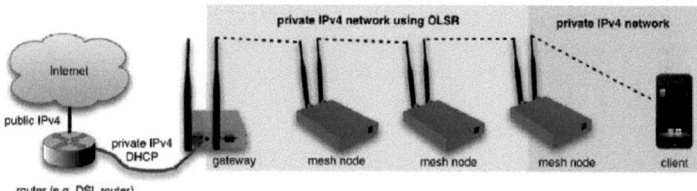

Figure 6.6: OViS network topology: OLSR (IPv4).

The finally selected third approach is depicted in Figure 6.6. It structures the network in three private IPv4 subnets, which are the router network, the WMN using OLSR for routing, and an access network for the OViS client. The access network is announced to the mesh nodes and the gateway by the *Host and Network Association (HNA)* mechanism of OLSR. The advantage of this approach is its simplicity. The OViS client can just use common IPv4 communication without any restrictions and special configuration. Thus, various platforms for the OViS client can be used. The network topology is completely transparent for the client. Potential drawbacks, such as special configuration of the last node to properly send HNA messages, can be completely hidden from the end user. Common video-conferencing solutions also hide the fact that all traffic from the OViS client has to pass through two nodes doing network address translation (NAT) and that the OViS client is not directly addressable from the Internet.

6.2.3 Multi-Channel Communication

The OViS network uses multi-channel communication to reduce the intra-flow interference and to increase the network throughput. The different links in the OViS

6.2. OViS CONCEPTS AND ARCHITECTURE

should not interfere with each other. Moreover, there might be several existing networks using 2.4 GHz in the surroundings of the construction site. Using IEEE 802.11a in the 5 GHz ISM band for the WMN, avoids the crowded 2.4 GHz ISM band with countless interfering nodes and provides more non-interfering/orthogonal channels for the multi-channel communication. However, as many smart phones and portable devices only support communication using IEEE 802.11b/g and we do not want to exclude the usage of these devices, the last link between the last mesh node and the OViS client still uses communication in the 2.4 GHz band.

During the network deployment, OViS has to automatically set up a multi-channel network using both frequency bands, 5 GHz and 2.4 GHz. Figure 6.7 explains the basic steps to connect an additional node to the wireless multi-channel network. The OViS client always uses channel 11 in the 2.4 GHz band to communicate with the node $N+1$ to be deployed in the next step. Node N is already deployed and running. As the next step, node $N+1$ is switched on. At startup, node $N+1$ sets its two wireless interfaces to channel 1 and 11 of the 2.4 GHz band. The OViS client can now communicate with node $N+1$ (see Figure 6.7a). In step 2, the OViS client reconfigures the node accordingly, i.e., the first interface is configured to use the same channel (here: X) in the 5 GHz as the second interface of the previously deployed node N via channel 11 (see Figure 6.7b). In step 3, the OViS client also reconfigures the second interface to a channel (here: Y) in the 5 GHz band. After this reconfiguration step, the OViS client cannot communicate with node $N+1$ anymore. Therefore, step 3 is omitted, if node $N+1$ is the last node to be deployed (see Figure 6.7c). The OViS client stores the state information about the channels used and is, therefore, able to appropriately distribute the channels during the whole deployment process. The channel assignment in OViS separates channels in order to avoid adjacent channel interference (ACI) and operates with the channel sequence 36, 104, 140, 40, 112, 48, 116, 52, 120, 56, 124, 60, 128, 64, 132, 100, 136.

6.2.4 Message Flow between OViS Client and the Mesh Node

The configuration and deployment process requires communication between the OViS client and the node that is currently deployed. This communication is always performed using channel 11 in the 2.4 GHz band. Figure 6.8 shows the message sequence of this communication. After being switched on by the user, a mesh node announces its presence to the OViS client by sending custom HELLO messages. The OViS client listens to these HELLO messages and then lists the detected nodes. The user now selects the current node, whose first interface is reconfigured afterwards. The OViS client node instructs the user to place the node according to the received RSSI measurements between it and the previous node. If the node has been deployed, the OViS client also reconfigures the second interface and instructs the user

6.2. OVIS CONCEPTS AND ARCHITECTURE

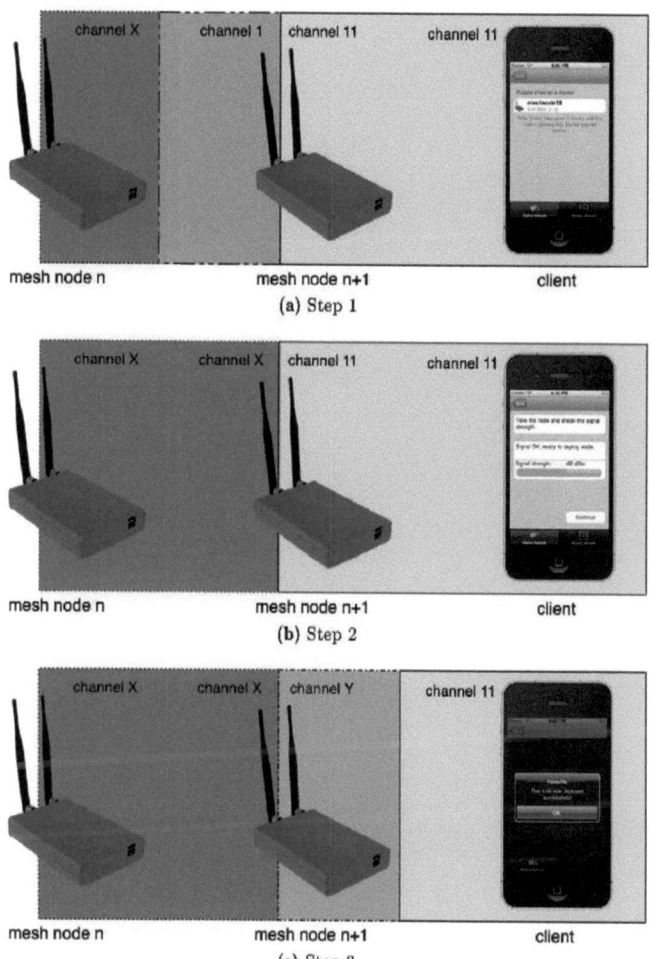

Figure 6.7: OViS: Deployment and configuration steps for multi-channel communication.

6.2. OVIS CONCEPTS AND ARCHITECTURE

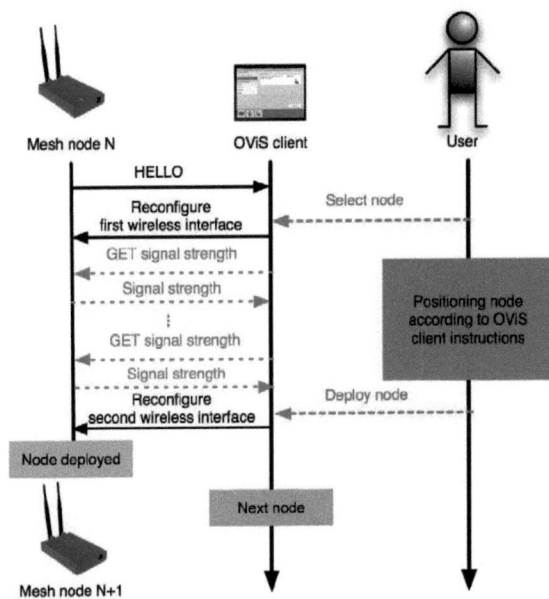

Figure 6.8: OViS: Message sequence for the deployment of a mesh node.

to take the next node.

It is worth noting that the entire management communication between the OViS client and the mesh node uses the permanently present IPv6 of ADAM (see Chapter 3). Management tasks are performed using this automatically configured IPv6 network, whereas data traffic uses freely configurable IPv4 or IPv6 networks. This approach provides the highest flexibility as it can be easily used today as all our target devices already include IPv4/IPv6 dual stacks. It is fully future-proof as it also works in a pure IPv6 network.

6.3 OViS Mesh Nodes

The OViS network consists of ordinary WMN nodes. Our prototype uses ALIX embedded boards (see Section 2.2.1) running our own embedded Linux distribution ADAM, described in Chapter 3. Besides the operating system tailored for WMN, OViS profits from ADAM's management feature and the single network configuration file per node. For communication, both wireless interfaces of the nodes are used in combination with the new wireless device driver *ath5k*. Lithium-Polymer batteries (12 V, 3200 mAh) are used to power the mesh node and to make them independent of the electric grid. Figure 6.9 shows our prototype hardware.

Figure 6.9: Prototype of a battery-powered OViS WMN node.

Figure 6.10 shows the main components of OViS on a mesh node. They consist of the *OViS pinger*, a web server, the *OViS signal strength monitor*, the *OViS network configurator*, and the *OViS network watcher*. The *OViS pinger* is responsible to announce the presence and the state of the mesh node to the OViS client. The HELLO messages are sent over UDP to the IPv6 link-local multicast address (*ff02::1*) and port 4379. They include host name and gateway status (yes or no) besides the source IPv6 address. The OViS client then uses this information to display

6.3. OVIS MESH NODES

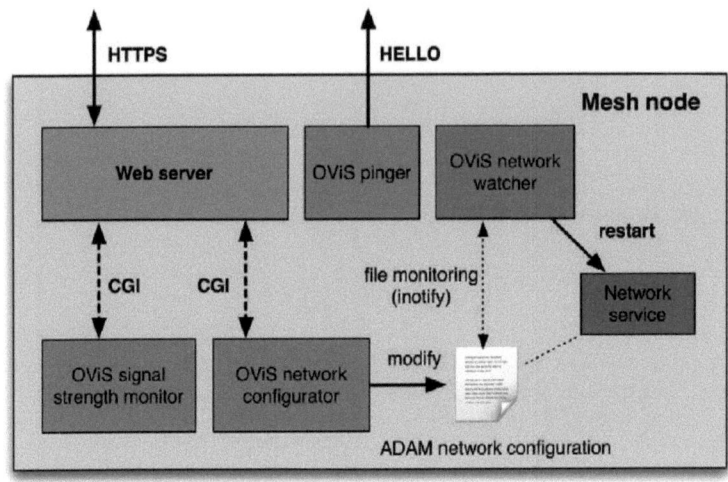

Figure 6.10: OViS components on a wireless mesh node.

the node and initiate communication for the deployment process. The node is configured through an HTTPS Common Gateway Interface (CGI). Using this interface, the client can retrieve the current RSSI values from the *OViS signal strength monitor* or remotely reconfigure the node. On request of the client, the *OViS signal strength monitor* delivers the current RSSI value, retrieved by the Linux standard tool *iw*. For reconfiguration, several commands are supported. The sequence *pass=kji9dkk, iface=wlan0, freq=5180, essid=ovismesh* reconfigures the first wireless interface of the node to use the frequency 5.180 GHz and to set the ESSID to one used for the mesh network. As only authorised clients should access the remote network configuration, an authentication token has to be delivered with every call. By using a CGI script for the implementation of the management communication, we profit from the socket management features and the traffic encryption of the existing web server. This significantly reduces the development effort. As CGI scripts are running with the unprivileged permissions of the web server, we cannot directly perform configuration actions. The *OViS network configurator*, therefore, only changes the *ADAM network configuration file* and cannot directly reconfigure the network interfaces. In order to immediately apply the modified network configuration file, we employ a new system service (*OViS network watcher*) that uses the new Linux file monitoring feature *inotify* to be notified in case of file modifications. If the file has changed, the *OViS network watcher* restarts the network service, which then applies the changes.

6.4 OViS Deployment Applications

The OViS client provides the interface between the user and OViS. It consists of two independent parts, the deployment wizard, and the video conferencing application. First, the OViS client has to guide the user through the entire deployment process. All technical details of the deployment have to be hidden from the user. The OViS client should only provide comprehensible instructions that any arbitrary person can follow in order to quickly deploy an OViS network and profit from it.

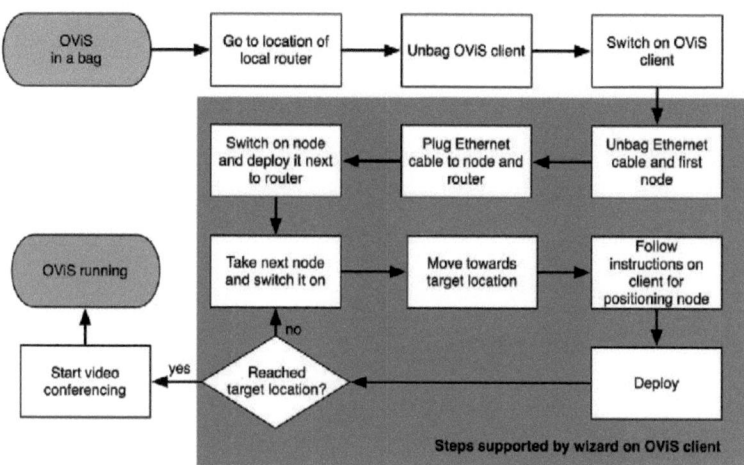

Figure 6.11: OViS Support Process: Deployment of a temporary communication infrastructure for on-site video conferencing by an inexperienced user.

Figure 6.11 illustrates the complete deployment process from the user's point of view. The user takes the bag with all the OViS components and first goes to the location of the local router, e.g., the on-site office in one of the construction containers. He/she takes out the OViS client, switches it on, and starts the OViS deployment application, which then guides the user throughout the deployment process. The user is first instructed to take the Ethernet cable and one of the nodes from the bag, to connect the node with the cable to the local router, and then to switch the node on. If the node works correctly and has network connectivity, the OViS client instructs the user to deploy it as a gateway. He/she is then instructed to unbag the next node, switch it on, and move with it towards the target location.

OViS now constantly monitors the RSSI values of the link to the previous node and provide feedback to the user. Therefore, the OViS client queries the current

6.4. OVIS DEPLOYMENT APPLICATIONS

RSSI value every second. If the RSSI value is too high, too many nodes might be required to reach the target location. The user is then instructed to proceed into the direction of the target. If the RSSI value is becoming too low, no constant network connection can be established. The user is then instructed to move back towards the previous node. According to our evaluations in Section 6.5, a reasonable RSSI value is between -50 and -70dBm for the used hardware. If the RSSI is in this range, the user is instructed to deploy the node and to take the next node. As soon as he/she has reached the target location, he/she stops the deployment and then starts the video conferencing system.

We developed OViS deployment applications for different software and hardware platforms. They include a command-line client, a personal computer (PC) client for Linux, Mac OS X and Windows operating systems, a full-screen kiosk application for an Ultra Mobile PC (UMPC), as well as smart phone applications for Android [84] and iOS [9] devices. Table 6.1 provides an overview on all developed OViS clients.

	Supported OS	Programming Language	GUI framework
Command-line	Linux, Mac OS X, Windows	Python	-
PC client	Linux, Mac OS X, Windows	Python	wxWidgets
UMPC client	Linux	Python	Tk/TkInter
Android client	Android	Java	Android
OViS App	iOS	Objective C	iOS

Table 6.1: Available OViS client applications.

The command-line client is mainly used for system tests (see Figure 6.12a). The PC client application for standard desktop and mobile computers has been written using the Python programming language and the cross-platform GUI library wxWidgets [161, 162]. It runs on Linux, Mac OS X and Windows operating systems and provides the operating system's native look and feel (see Figure 6.12b).

For our target scenario, the construction site, the command-line and the PC client are not suitable. An integrated appliance with special hardware is required. Therefore, we developed a prototype using an Asus R2H Ultra Mobile PC (UMPC), a mobile computer with a touch screen (800 x 480 pixels). It contains a 900 MHz Intel Celeron CPU, 1.2 GB RAM and a wireless network interface and runs a Linux operating system. The UMPC client is developed with Python programming language and the the GUI library Tk [58, 141]. In order to provide a smooth user experience of a highly integrated appliance, the UMPC automatically logs a specific user in and starts the OViS application. After the successful deployment, the

6.4. OVIS DEPLOYMENT APPLICATIONS

(a) OViS command-line client

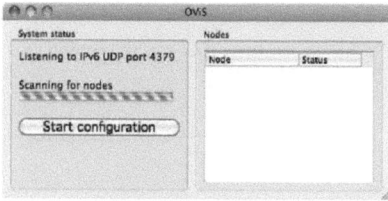

(b) Mac OS X OViS client

Figure 6.12: Command-line and graphical OViS clients for personl computers.

user can directly start the video conference (e.g., Skype) from the application. Figure 6.13 shows the UMPC full-screen *kiosk* application. It guides the user through the deployment process as described earlier. Besides textual instructions, it uses a graphical feedback for the node placement. If the RSSI value is around the ideal value of -60 dBm (± 10 dBm), two green arrows indicate that the node should be placed here. If RSSI value is between ± 10 dBm and ± 20 dBm of the ideal value, yellow arrows indicate if the user have to move either closer or further away from the previous node. Red arrows are used if the value varies more than ± 20 dBm. Additionally, an acoustic feedback has been introduced such that the user does not need to look constantly on the screen. The repetition speed of the beeps maps to the colours of the graphical representation, i.e., no beeps = green, slow beeps = yellow, fast beeps = red. The pitch of the acoustic beep indicates the direction. Beeps with low pitch instructs the user to move further away, whereas beeps with high pitch instruct the user to move closer to the previous node.

6.4. OVIS DEPLOYMENT APPLICATIONS

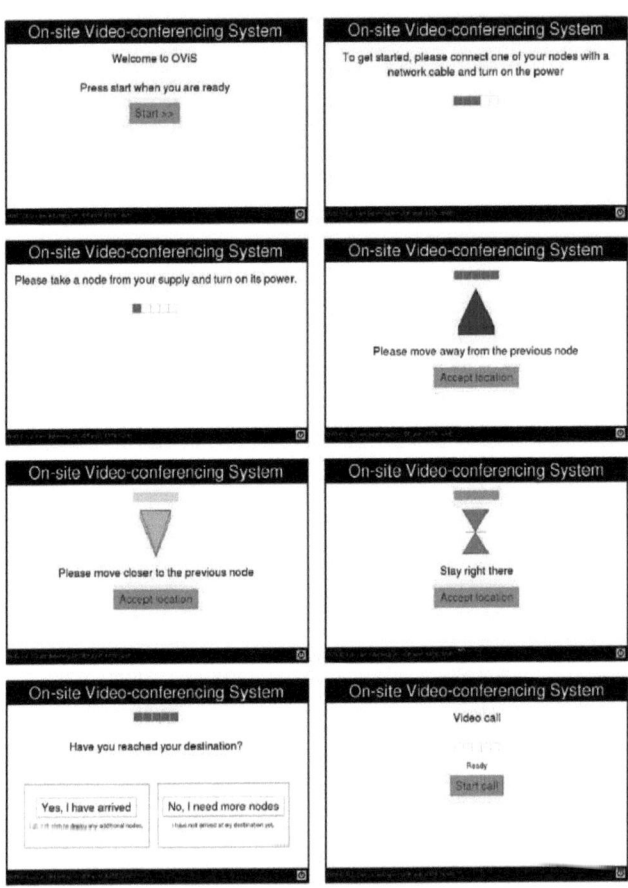

Figure 6.13: OViS full-screen *kiosk* application optimised for the Asus R2H UMPC.

6.4. OVIS DEPLOYMENT APPLICATIONS

Figure 6.14: OViS deployment process guided by an Android application (Part I).

6.4. OVIS DEPLOYMENT APPLICATIONS

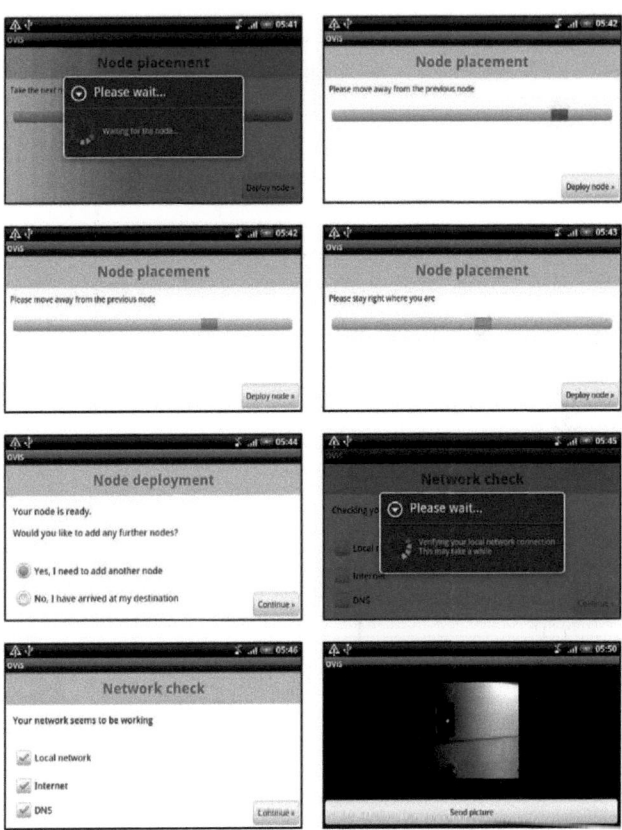

Figure 6.15: OViS deployment process guided by an Android application (Part II).

6.4. OVIS DEPLOYMENT APPLICATIONS

Figure 6.16: OViS deployment application for iOS (running on iPhone).

6.4. OVIS DEPLOYMENT APPLICATIONS

Using the heavy UMPC is still not the best solution; therefore, we implemented OViS clients for the smart phones based on the two major platforms Android and iOS. Android is a Linux based operating system developed by Google. It runs on smart phones from different vendors. Apple's iOS is the operating system of the widely spread iPhones and iPads. Although, there are some problems, e.g., lacking ad-hoc networking support of the used Android version, the smart phone applications represent the best user experience of OViS. As the Android phone used for development does not have a front-facing video camera, the current Android application only provides a SIP based audio communication and the exchange of static pictures. Screenshots are shown in Figures 6.14 and 6.15. In order to support all currently mobile devices of Apple, we developed a universal application that runs on iPod touch, iPhone and iPad. The GUI is adapted to the capabilities of the devices (see Figures 6.16 and 6.17). For video conferencing, existing applications, such as Skype or Apple's Face-time, can be used.

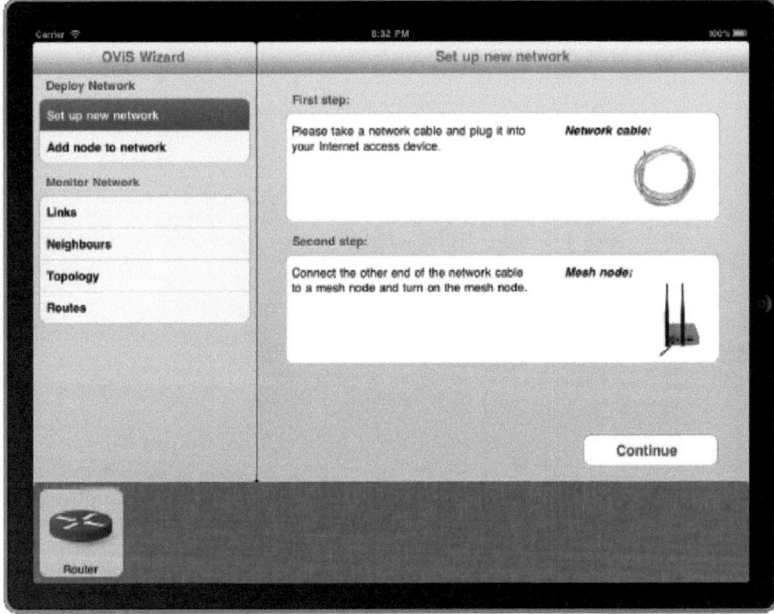

Figure 6.17: OViS deployment application for iOS (running on iPad).

6.5 Evaluation

The evaluation of OViS compromises two parts: the determination of RSSI thresholds and the actual performance evaluation.

6.5.1 Determination of RSSI Thresholds

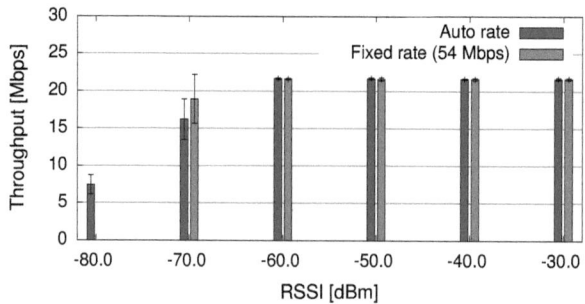

Figure 6.18: Achievable single hop throughput in relation with the received signal strength indicator (RSSI).

Before evaluating the performance of OViS, we have to determine reasonable RSSI threshold values to deploy the mesh nodes. It is a trade-off between the minimum number of nodes to bridge a distance and required network bandwidth to support a high quality video conference. Therefore, we evaluated the achieveable TCP throughput versus the RSSI values over a single hop using NetPIPE [163, 164]. Figure 6.18 shows the measured maximum achievable TCP throughput in relation to RSSI values, when either using automatic data rate control or the fixed data rate of 54 Mbps. With the used hardware and automatic data rate control, a connection can be established if the RSSI value is at least -85 dBm, although the connection is very unstable. Stable connections can be established if the RSSI value is higher than -80 dBm resulting in a network throughput higher than 7 Mbps. The maximum throughput significantly augments by increasing RSSI values, e.g., 16 Mbps for -70 dBm and 22 Mbps for -60 dBm. The maximum throughput is reached at an RSSI value of -60 dBm, i.e., higher RSSI values do not result in any improvements. When using a fixed data rate of 54 Mbps instead of the automatic data rate control,

6.5. EVALUATION

the throughput is increased from 16 Mbps to 19 Mbps for an RSSI value of -70 dBm. However, for values below -70 dBm, connections are unstable. Therefore, we selected the automatic data rate control mechanism due to the increased stable operation domain (up to -80 dBm). We have set the target RSSI threshold for the deployment of an OViS node to -60 dBm (±10 dBm). The tolerance range has been defined to avoid contradicting instructions to move back and forth due to RSSI fluctuations resulting from the user moving in walking pace. The target RSSI value of -60 dBm includes a tolerance margin and ensures stable network operation even in case of temporary varying interfence conditions. The deployed network then sustains video conference at a decent quality in any circumstance. Therefore, we configured the OViS clients to check the RSSI every second and to instruct the user to deploy the mesh node if two consecutive RSSI values of -60 dBm (±10 dBm) are measured on the link to the previous node. As a node is usually carried at around 1 m above ground and then perhaps deployed directly on the ground, the actual RSSI after the deployment might be worse.

6.5.2 OViS Performance Evaluation

After a successful determination of reasonable RSSI thresholds for the OViS deployment, we started our evaluation of OViS by verifying its usability and performance in a real world situation. Figure 6.19 depicts the scenario for the test deployment. A temporary network consisting of four mesh nodes over three floors is deployed. One notebook is deployed next to the first node on the top floor. The three other nodes are then either placed "randomly" by an inexperienced user or deployed according to the instructions of the OViS deployment wizard. The TCP throughput was measured between the two notebooks using NetPIPE [163, 164].

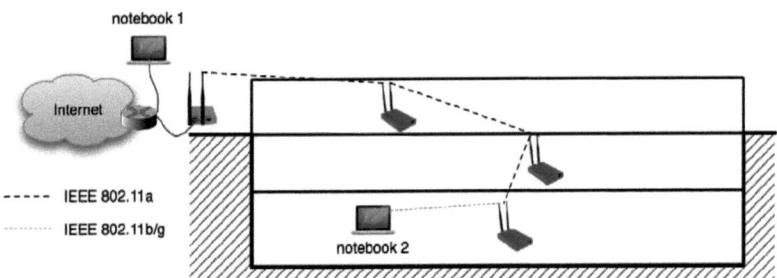

Figure 6.19: OViS deployment test scenario.

Figure 6.20 shows the TCP throughput measured on the network deployed with

158

6.5. EVALUATION

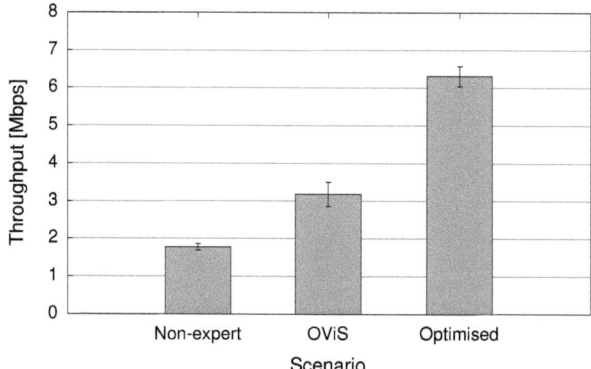

Figure 6.20: Throughput of different deployments: non-guided deployment, OViS deployment, and manually optimised.

and without using the instructions of the OViS client. In addition, it shows the results of a manually optimised deployment. The values represent the average network throughput and the standard deviation. The OViS network reaches a throughput of 3-3.5 Mbps, which is sufficient for a video-conferencing system. In order to compare the performance of the OViS deployment with the performance of a non-guided (random) deployment, a non-expert user was instructed to just deploy three intermediate nodes at his discretion. This deployment only offered a throughput of 1.8 Mbps, the half throughput of the OViS network. Depending on the deployment choice of the user, the throughput could be either lower or higher and reaching the performance of OViS. The user could even select an insufficient number of nodes resulting in a non-working network. For example, network connectivity could not be reached with only one intermediate node in our scenario whereas two intermediate nodes would be sufficient. In order to classify the effectiveness of OViS, we manually optimised each link of the network deployed by OViS. By realigning the antennas and moving the nodes a few centimetres, we improved the RSSI values of all links still covering the same distance. This optimisation took more than 15 min and was only valid at this point in time. Therefore, it only serves as optimal value. Figure 6.21 shows the improved RSSI values compared with the original ones, measured with the Linux standard tool *iw*. The values represent the average and the standard deviation for ten measurements. Our fine-tuning improves the RSSI values by about 10 dBm on the vertical links. This is mainly due to antenna adjustments. The used antennas have a higher gain in the horizontal plane. After the manual adjustments, the

6.5. EVALUATION

throughput was measured again and a significant improvement of the throughput to values of 6-6.5 Mbps was observed. A proposal for getting better results when using OViS is to extend our nodes with two additional antennas. Using one vertically aligned antenna and one horizontally aligned antenna per wireless interface and enabling antenna diversity support, OViS could provide better results on the vertical distances.

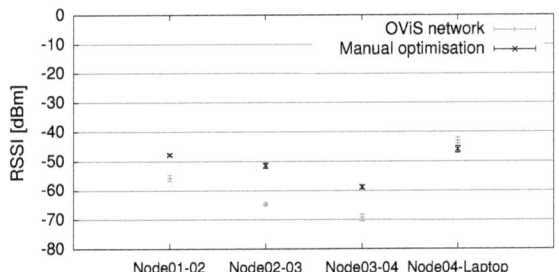

Figure 6.21: Signal strengths achieved after deployment with OViS and after manual optimisation with relocating the nodes and aligning the antennas.

In summary, the results show that OViS can guide an inexperienced user in the deployment of a temporary network that meets the requirements of a reliable video conference. It provides a significantly better result than a non-guided deployment and guarantees network connectivity. The average network throughput is more than sufficient for the use case. However, the measurements after the manual fine-tuning show that the OViS deployment is not perfect, but this is not necessary to meet the communication requirements. An improvement for the throughput of the OViS deployment may be achieved by using multiple antennas with antenna diversity. The currently used antennas are slightly directed in the horizontal plane, an additional vertically directed antenna might be beneficial.

6.5.3 Multi-Hop Throughput

Although the single hop measurements showed a maximum throughput of 22 Mbps in Figure 6.18, the optimised multi-hop deployment only offered a throughput of 6-6.5 Mbps. Therefore, we evaluated the throughput with an increased hop count. The test scenario was as follows: The network first consisted of two nodes com-

6.5. EVALUATION

municating with each other over a wireless link (IEEE 802.11a). These two nodes were connected by Ethernet to the notebooks running NetPIPE. Intermediate nodes were then added to the setup, forming a chain topology by using the channels 36 (5180 MHz), 104 (5520 MHz) and 140 (5700 MHz) for the individual links. Figure 6.22 shows the measured average throughput values as well as the standard deviations for a fixed rate of 54 Mbps and automatic rate control. A single link offers a throughput of 21 Mbps respectively 21.5 Mbps, as already measured in Figure 6.18. Although orthogonal channels were used, the throughput is drastically decreased for two (6.4 Mbps respectively 12 Mbps) and three hops (5.6 Mbps respectively 7.5 Mbps). The measurements are inline with Figure 6.20. The multi-channel communication did not provide the expected multi-hop throughput. Therefore, we quantified the effect of multi-channel communication in the next experiment.

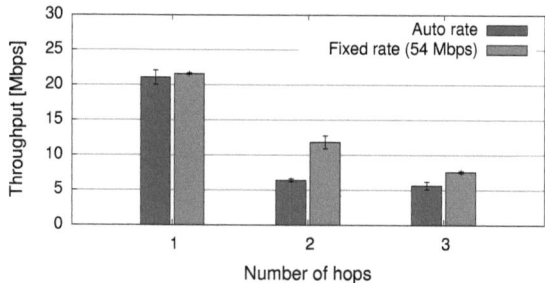

Figure 6.22: Throughput depending on the number of hops.

6.5.4 Multi-Channel Performance

The following evaluation quantifies the benefit of using multi-channel communication in OViS. The test scenario consisted of a network with three nodes placed in a chain topology. All links provided a signal quality of at least -50 dBm. First, all nodes were communicating on the same channel 36 (5180 MHz). Second, the two links used the channels 36 (5180 MHz) and 104 (5520 MHz). Figure 6.23 depicts the received results. In the used scenario, multi-channel communication improved the throughput by 1 Mbps due to the reduced interference between the two links. Multi-channel communication is, therefore, beneficial, although it did not delivered the expected benefits. Possible reasons are adjacent channel interference [47, 49],

6.6. CONCLUSIONS

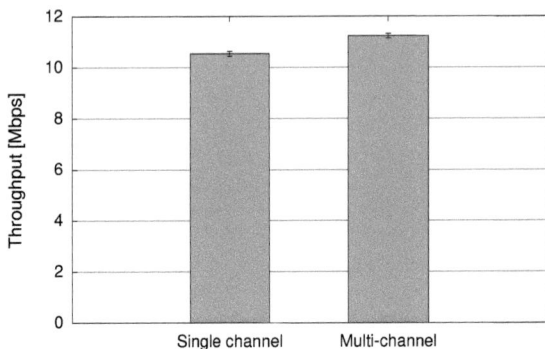

Figure 6.23: Throughput in a two hop scenario using single and multi-channel communication.

board crosstalk and radio leakage of the wireless cards [131], or insufficient physical separation of two antennas [7].

6.6 Conclusions

In this chapter, we presented OViS as a support framework for an ad-hoc deployment of WMNs. OViS targets the use case of a temporary communication infrastructure for an audio/video conferencing system on a construction site to reduce costs. By using the audio/video conferencing system, an electrical engineer can support multiple construction sites more efficiently. Unfortunately, most problems usually occur at the switching units in the basements of the building in constructions without any network coverage by wired or cellular networks. OViS provides network connectivity by deploying a temporary battery-powered WMN, which propagates the Internet access from the on-site office to the basements. Employing a wireless network, OViS does not introduce additional safety risks, such as tripwires. However, the deployment of a WMN is usually a task for a network expert. The contribution of OViS is to enable an inexperienced user to easily and quickly deploy the WMN. OViS achieves this by automatically configuring the network and by a deployment wizard (application) that guides the user with easily understandable instructions to place the WMN nodes in appropriate distances.

We developed OViS clients for the following platforms:

- Personal computers with Linux, Mac OS X or Windows operating systems

6.6. CONCLUSIONS

- OViS integrated appliance based on an UMPC with a touchscreen
- Smart phones and mobile devices with Android
- iOS based devices such as iPod touch, iPhone and iPad

Using one of these OViS clients, an inexperienced user can quickly deploy a working communication infrastructure for an audio/video conferencing system, which has been verified in our evaluation. A network deployed according to the instructions of the OViS client is always connected and provides a throughput of 3-4 Mbps, which is more than sufficient for the use case.

The implementation of OViS was simplified by the ADAM build system and the ADAM management framework. OViS run the embedded Linux distribution created with the ADAM build system on its WMN nodes. Due to the modularity of the build system, the newly developed software for the OViS nodes, e.g., the OViS pinger, as well as all necessary extensions of existing software could be easily integrated. The entire process of building the firmware for the OViS has been automated by employing the ADAM build system. In addition, ADAM's concept of one single network configuration file per node simplified the re-configuration process used in OViS.

Our measurements showed that used automatic rate control mechanism provides a more reliable network connection, but a lower throughput than just fixing the data rate to the maximum. A better data rate control mechanism would be beneficial. The multi-hop performance needs to be further investigated. Instead of using a fixed channel allocation, OViS could make use of a multi-channel MAC protocol. The OViS network could then automatically adapt the used communication channels to the currently received interference. Extending the OViS with a third wireless interface dedicated for the client access would be beneficial to allow adding nodes after the first deployment and to cover areas instead of just bridging a distance.

In the next chapter, we further automatise the network deployment and investigate a completely autonomous network deployment using flying robots.

Chapter 7
Autonomous Deployment of a Wireless Mesh Network using Unmanned Aerial Vehicles

Whereas a semi-automatic (guided) deployment of a temporary WMN by non-expert users was described in Chapter 6, this chapter introduces a framework for completely autonomous deployment of WMNs [87, 129]. The framework, termed UAVNet, focuses on a deployment of a WMN using small Unmanned Aerial Vehicles (UAVs). There are several application scenarios where a deployment of a communication network is beneficial. A major application of UAVNet is to replace destroyed or missing communication infrastructure, e.g., in emergency or disaster recovery scenarios. UAVNet enables connectivity between end systems of rescuers using one or more flying mesh nodes. Extensions to provide network coverage over a pre-defined area are possible. Other applications are the support for environmental monitoring or data aggregation in remote wireless sensor networks for agricultural purposes.

UAVNet includes a concept and a prototype implementation of an autonomously deployable temporary WMN consisting of WMN nodes carried by small quadrocopter UAVs. The flight control electronics of these UAVs is connected to the WMN nodes over a serial line (see Figure 7.4). Hence, a custom UAV controller service on the WMN node can indirectly control the flight of the UAV by adding/removing navigation points. The prototype of UAVNet can autonomously interconnect two communication peers (clients) by establishing an aerial WMN between them.

The structure of this chapter is as follows: Section 7.1 discusses the motivation for the development of UAVNet. After the presentation of the considered scenarios in Section 7.2, the components of UAVNet and the communication protocol between these components are shown in Section 7.3 and Section 7.4 respectively. Section 7.5 presents our remote control application for iPhone/iPad. In Section 7.6, we provide an evaluation of our prototype. Section 7.7 concludes this chapter.

7.1 Introduction

In first response and disaster recovery scenarios, e.g., after avalanches, flooding, or earthquakes, communication infrastructures are often not available as they have not been previously deployed in the affected area or have been destroyed during such an event. As efficient distribution of information among the rescuers and the tactical operation centre is crucial, a temporary communication infrastructure is set up in the first phase of a rescue operation. This temporary communication network enables multimedia communication between the rescuers and the tactical operation centre. Multimedia data consisting of pictures and videos may help in the assessment of the situation and finally in delivering a common situation report that facilitates operation control.

A WMN can provide such a temporary broadband communication infrastructure for emergency and disaster recovery scenarios. However, as it has to be deployed immediately after the start of the rescue operation, the deployment should be easy, requiring only limited man power. The network should be constantly adaptable to the communication needs. An ideal solution is a fully autonomous network deployment without manual interaction.

There are two options for the autonomous deployment of WMN nodes. The nodes can be carried by either land-robots or UAVs (flying robots). As traversing terrain following a natural disaster can be very challenging, land-robots face several mechanical problems, e.g., climbing an obstacle. Some sites might even not be reachable by a land-robot.

UAVs offer a solution to avoid these mechanical challenges and provide better site accessibility. In addition, they could provide better network coverage (air relays) than land robots. As first response scenarios are usually very dynamic, a flying WMN offers better adaptability to the current communication needs of the rescuers. The UAVs can further deliver aerial images of the event. Therefore, we selected UAVs as platform for autonomous deployment of temporary WMNs for emergency operations, called UAVNet.

A drawback of UAVs is their continuous energy consumption needed to stay airborne in contrast to land-robots, which can go to sleep as soon as they have reached its final destination. A complete solution, therefore, requires strategies for autonomous replacement and recharging of the UAVs.

UAVNet can provide broadband network connectivity by a flying WMN. It combines existing WMN technology with an existing quadrocopter UAV platform to provide an ad-hoc communication infrastructure, which is established autonomously and can dynamically adapts to the current communication needs. Thus, each UAV carries a small WMN node, which is interconnected to its flight electronics in order to adapt the flight according to the deployment scenario, i.e., a controller component on the WMN node can adapt the current flight of the UAV. Multiple communicating UAVs then form the flying WMN, such as the swarm shown in Figure 7.1. The WMN is automatically established using IEEE 802.11s (see Section 2.1.1).

Figure 7.1: Flying UAV swarm carrying a temporary WMN.

7.2 Scenario

The current version of UAVNet is able to establish ad-hoc WMNs for one specific scenario, namely the airborne relay scenario. In this scenario, a temporary WMN is established to interconnect two distant clients. The network consists of one or multiple flying communication relays (see Figure 7.2) and a remote control client providing a user-friendly control interface to initialise and monitor the network deployment.

7.2.1 Search Mode

We assume that the location of at least one client is known when deploying a temporary WMN with UAVNet. In order to find the second client, either the user provides UAVNet the direction towards the approximate location of the second client or he/she instructs UAVNet to perform an autonomous search. In the first case, i.e., given an approximate location, the UAV first flies into the specified direction until it discovers the second client by receiving a position update. The second client is

7.2. SCENARIO

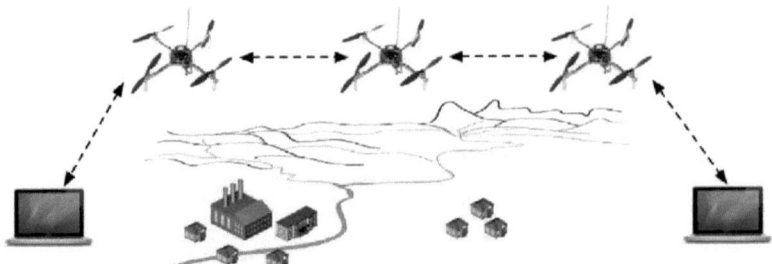

Figure 7.2: Multi-Hop Airborne Relay Scenario.

identified by its MAC address. In case that multiple clients are present, client identification is assisted by a predefined list. In the second case, when using autonomous search mode, the UAV flies on a spiral track around the first client in order to find the second client.

7.2.2 Positioning of UAVs

UAVNet includes two positioning options for placing the UAVs between the distant clients. It can position them using either GPS coordinates of the clients or signal strength measurements in addition to GPS coordinates. If only GPS coordinates of the clients are used, the UAVs are placed evenly on the connection line between the two clients. An obvious drawback of this positioning approach is that it does not consider the variation of client-specific communication ranges nor any environmental interference resulting in varying connection quality.

An improved positioning approach is to place the UAVs according to the received signal strengths. This positioning process is explained for a *single airborne relay* in Section 7.2.3 and for a *multi-hop airborne relay* in Section 7.2.4.

7.2.3 Single Airborne Relay

In the *single airborne relay* scenario, the UAV is first placed on the connection line equidistant between the clients. Then it measures the signal strengths from both clients and flies towards the client with the lower signal strength. During its repositioning, the UAV constantly monitors the received signal strengths from both clients. The positioning process completes when the signal strengths to both clients are equal. A threshold is used to avoid permanent repositioning.

Figure 7.3 shows the process of connecting two distant clients by a single airborne wireless relay. The two clients (*client 1* and *client 2*) are running but out of

7.2. SCENARIO

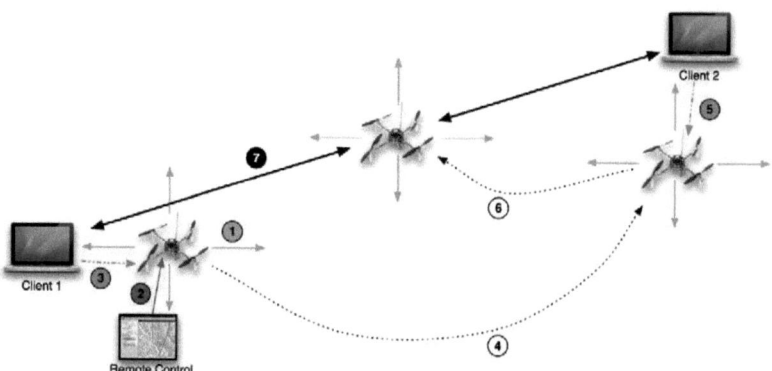

Figure 7.3: Process of connecting two distant clients by one single flying WMN node (airborne relay).

communication range of each other. The deployment process of UAVNet consists of seven numbered steps:

1. The deployment of UAVNet begins with the first UAV. When being switched on, a UAV permanently announces its presence.

2. A remote control client, listening to these announcements, discovers the UAV. The user selects the UAV and the appropriate deployment scenario (here: *single airborne relay*) on the remote control client, which then instructs the UAV to start the network deployment.

3. The co-located *client 1* also receives the presence announcements of the UAV and replies by providing its own geographic position to the UAV.

4. The UAV now searches for the other client either by flying into a predefined direction or by an autonomous search.

5. As soon as the UAV comes into communication range of *client 2*, its presence announcements trigger the transmission of the geographical position of *client 2*.

6. The UAV now positions itself in-between the two clients and then flies towards the client with the lower signal strength until the signal strengths to both clients are equal.

7. The two clients can now communicate using the established airborne relay.

7.2. SCENARIO

7.2.4 Multi-Hop Airborne Relay

In case of a *multi-hop airborne relay* scenario, the network deployment employs first the same steps as the single airborne relay (steps 1-5) to determine the position of *client 2*. The only difference is that the user has to select the option *multi-hop airborne relay* on the remote control client at the initialisation of the network deployment. After having completed the first five steps of the single airborne relay deployment, the deployment of multi-hop airborne continues with steps 6 - 14:

6. After discovery of *client 2*, *UAV 1* positions itself in the middle of the connection line between the two clients.

7. Afterwards, it flies towards *client 1* until the measured received signal strength from the first client reaches a pre-defined value.

8. *UAV 1* announces that it has reached its final position by notifications to the remote control client.

9. *UAV N+1* is switched on and receives the positions of both clients and all previously deployed UAVs from the remote control client.

10. *UAV N+1* is first positioned 10 m away of *UAV N* into the direction the second client. As it knows all positions of the previously deployed UAVs, collision avoidance is simple as *UAV N+1* can simply fly around the already deployed UAVs.

11. *UAV N+1* then continues to fly into the direction of the second client and constantly monitors the received signal strength to *UAV N*. If this received signal strength matches a pre-defined value, *UAV N+1* stays at its current position.

12. *UAV N+1* announces its arrival at the final destination to the remote control client.

13. Steps 9-12 are repeated until the currently deployed UAV has a sufficient connection to the second client, i.e., the received signal strength matches a pre-defined threshold.

14. The two clients can now communicate using the established multi-hop airborne relay.

Table 7.1 summarises the currently supported deployment scenarios and options, which a user can selected on the remote control client. In addition to the eight possible combinations of the airborne relay scenario, a monitoring option offers monitoring of the current network deployment by multiple remote clients.

	Search mode	Positioning
Single airborne relay	directional or autonomous	geographically or signal strength
Multi-hop airborne relay	directional or autonomous	geographically or signal strength
Monitoring	-	-

Table 7.1: UAVNet deployment scenarios and options.

7.3 System Components

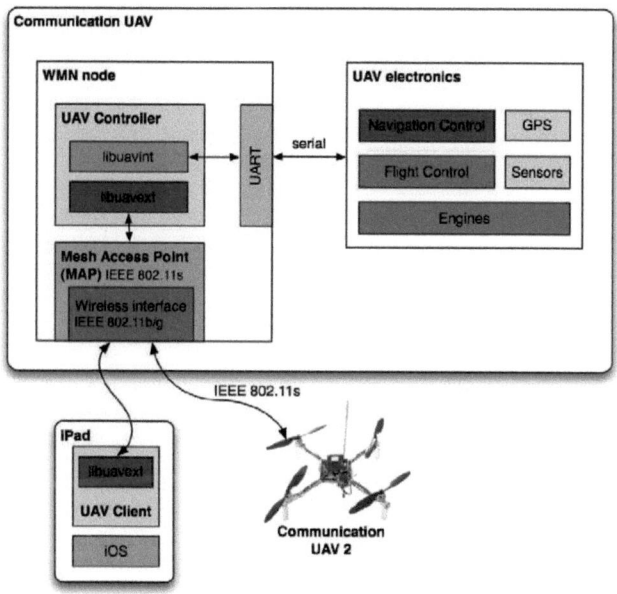

Figure 7.4: System components of UAVNet: WMN node with UAV controller and IEEE 802.11s mesh access point (MAP), UAV electronics and UAV client.

Figure 7.4 presents the system components of UAVNet belonging to the communication UAVs and to the remote control units. A communication UAV consists of the UAV electronics and a WMN node. The UAV electronics contain all hardware components and software necessary to control the flight of the UAV. The attached

7.3. SYSTEM COMPONENTS

mesh node can monitor and control the flight of the UAV using a serial connection to the *Navigation Control*. As a main logical component, UAVNet introduces the *UAV Controller*, which runs on the mesh node. Its purpose is to control the flight of the UAV towards a specific location, according to the current scenario. It is, therefore, responsible for all communication between the mesh node and the UAV electronics. By implementing the serial protocol of the used UAV platform [36], the *UAV Controller* can control the flight path of the UAV by adding and removing navigation points in the flight list of the *Navigation Control*. Thus, the *UAV Controller* can instruct the UAV to fly to a specific location. Moreover, the *UAV Controller* processes the periodic announcements of the *Navigation Control*, containing important flight parameters such as the current GPS position, height, flight direction, speed and battery level. All functionality for communicating with the UAV electronics is encapsulated in the library *libuavint*.

For flight coordination, scenario control and monitoring, the *UAV Controller* has to communicate with other UAVs and the remote control clients. For this, it uses the WMN established by IEEE 802.11s (see Section 2.1.1). The entire functionality to handle the external control traffic is encapsulated in the library *libuavext*, which is available for different UNIX based operating systems, e.g., Linux, MacOSX and FreeBSD. The different control messages are described in detail in Section 7.4.1.

The remote control client, in our case an iPhone/iPad, uses the library *libuavext* to control and monitor the UAVNet WMN using IEEE 802.11s connections. It offers the user a convenient interface to select the deployment scenario of UAVNet. Afterwards, it provides monitoring capabilities by subscribing to the notification service running on the WMN nodes.

7.3.1 Communication Types

Figure 7.5 depicts all employed communication types in UAVNet. Serial communication is used to interconnect the on-board WMN nodes to the flight electronics. IEEE 802.11s is used for the control traffic between the individual UAVs with attached WMN nodes, and between the WMN nodes and remote control client, i.e., for exchanging scenario control and notification messages. The established IEEE 802.11s WMN is used for the data communication between the network clients.

7.3.2 Prototype

For our prototype, we use small quadrocopter UAVs with on-board WMN nodes. The quadrocopters are based on the Mikrokopter platform (see Section 2.6). The quadrocopters run the standard unmodified firmware from Mikrokopter community project. In order to implement the UAVNet prototype, it has not been necessary

7.3. SYSTEM COMPONENTS

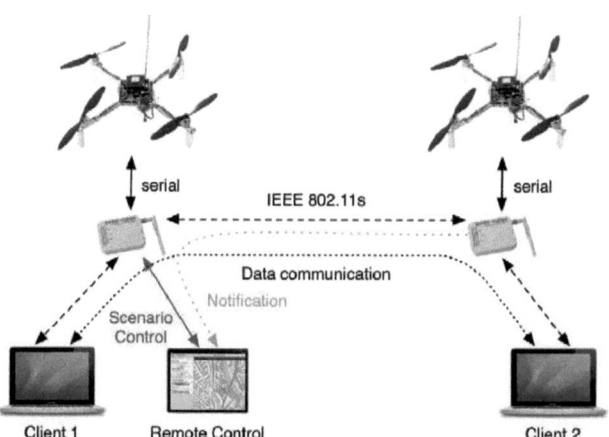

Figure 7.5: Communication types in UAVNet: Serial to interconnect WMN node and UAV, IEEE 802.11s for data and control traffic.

to adapt the firmware of the UAV. The complete UAVNet software has been implemented on the on-board WMN node, which is an OpenMesh OM1P node (see Section 2.2.1). Figure 7.6 shows the flight electronics and the on-board WMN node. The serial port of the OM1P is connected to the debug port of the Navigation Control of the flight electronics through a logic level converter (3.3 V to 5 V).

The WMN nodes run our own embedded Linux distribution ADAM as operating system (see Chapter 3). The ADAM build system is an ideal fit for the development process of the UAVNet software by providing an easily understandable and usable cross-compilation build system. It offers the same software on all supported hardware platforms, including the necessary networking components for UAVNet and full IPv4/IPv6 support. Routing is performed on layer 2 by the IEEE 802.11s (see Section 2.1.1). The recent Linux kernel versions in ADAM support the up-coming standard IEEE 802.11s by the open source implementation open80211s [137]. The support for the special wireless chip set of the OM1P node has to be added to the ath5k wireless driver [69].

Figure 7.7 shows a flying UAV with an on-board WMN node. Our first prototype of UAVNet consists of three of these UAVs, two notebooks as clients, and an iPhone or iPad as remote control client, which is described in more detail in Section 7.5.

7.3. SYSTEM COMPONENTS

Figure 7.6: UAVNet: Flight electronics connected by a serial connection to the WMN node.

Figure 7.7: Flying quadrocopter UAV carrying a WMN node.

7.4 Communication Protocol

UAVNet requires a communication protocol to coordinate the network deployment between all network participants, such as UAVs, remote control clients, and WMN clients. For example, a UAV has to announce its presence, the remote control client has to send commands to a UAV, and the clients have to submit their positions.

7.4.1 Protocol Messages

Figure 7.8 shows the UAVNet protocol messages that are exchanged between UAVs, clients and the remote control application. There are eight message types of UAVNet necessary to coordinate the autonomous network deployment: *HELLO*, *SCENARIO*, *OWN POSITION*, *ACK*, *ABORT*, *DIRECTION*, *UN-/SUBSCRIBE*, and *NOTIFICATION*.

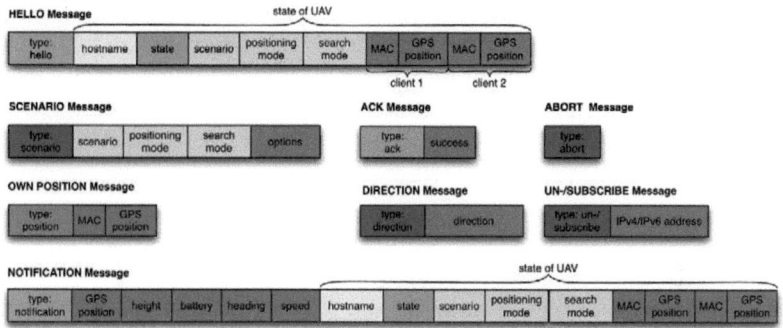

Figure 7.8: UAVNet: Protocol messages.

- *HELLO* messages are periodically broadcasted by a UAV to announce its presence. They contain the host name of the UAV (e.g., *uav01.unibe.ch*) and the current state including the running scenario (e.g., airborne relay), positioning mode (geographic coordinates or signal strength), search mode (direction or search spiral) and, if available, MAC addresses and GPS positions of the clients.

- A *SCENARIO* control message is transmitted by the remote control client to a UAV in order to start the network deployment. It contains the selected scenario, the selected positioning mode, the selected search mode, and additional options, such as a list of the client MAC addresses.

7.4. COMMUNICATION PROTOCOL

- An *OWN POSITION* message contains the geographic coordinates of a network client.

- An *ACK* message provides feedback on the correct application of a control message, e.g., a *SCENARIO* message.

- An *ABORT* message is used to abort the network deployment.

- A *SUBSCRIBE* message contains the IPv4/IPv6 address of a subscriber for the notification service of a UAV.

- A *NOTIFICATION* message is periodically sent to all subscribers of the notification service on a UAV. It includes the current position, height, battery level, heading, speed, host name, and current state of the UAV. Notification messages are routed within the WMN whereas *HELLO* messages are link-local.

The UAVNet messages are transmitted either over TCP or UDP. The reliable transport service of TCP is used for all control messages, namely *SCENARIO*, *OWN POSITION*, *ACK*, *ABORT*, *DIRECTION*, and *SUBSCRIBE* messages, as the loss of one of these message would affect the control flow. The UAV is listening for these messages on port 7654. The *HELLO* messages are broadcasted and thus use UDP to port 7655. *NOTIFICATION* messages are sent as unicast messages also using UDP but on port 7656.

7.4.2 Message Flow

In the following, the three message flows of the UAVNet communication during the network deployment are explained in detail. There are the two message flows for the deployment scenarios with and without a given direction to the second client and the message flow for the notification service.

Figure 7.9 depicts the combined message flow of the two deployment scenarios, a) with the manual search and b) with the autonomous search mode. When using option a), the direction towards the location of the second client is known and can, therefore, be delivered by the remote control client. The user switches on the remote control client. Then he/she switches on the first UAV, which immediately announces its presence by the periodic broadcast of *HELLO* messages. The remote control client discovers the UAV allowing the user to configure the deployment scenario, e.g., the single hop airborne relay. The user determines the direction of the approximate location of the second client and selects either positioning by geographic client locations or by signal strength measurements. The remote control client then starts the deployment process by sending a *SCENARIO* control message to the UAV.

7.4. COMMUNICATION PROTOCOL

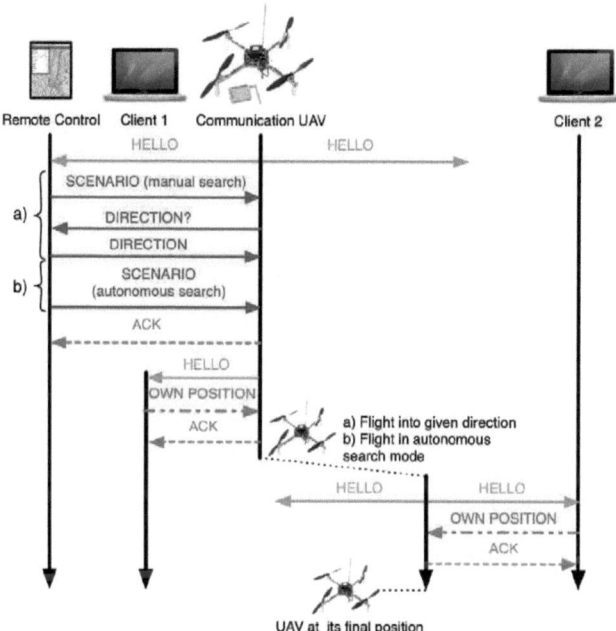

Figure 7.9: Message flow for a scenario with and without known direction towards the location of the second client (manual and autonomous search).

7.4. COMMUNICATION PROTOCOL

As option a) with the manual search has been selected, the UAV requests the direction from the remote client, which delivers it by transmitting a *DIRECTION* message. The UAV acknowledges the correct reception of the *DIRECTION* message. Concurrently, the co-located *client 1* replies to the *HELLO* messages by sending its own position. If the UAV has received the position of the first client, it sends an acknowledgement and then starts its flight into the specified direction towards *client 2*. If the UAV is within the communication range and the state in the *HELLO* message requests a position of a client, *client 2* replies to a *HELLO* message by sending its own position. Having the positions of both clients, the UAV can now fly to its final position either by using geographic positioning or by signal strength measurements. After establishing the WMN, the two clients are able to communicate over the airborne relay.

In case of an unknown position of the second client, the message flow with option b) is used (see Figure 7.9). It is almost the same as in the first scenario, but without transmitting a *DIRECTION* message. After having all necessary settings, the UAV then searches the second client by flying on a spiral track around the first client. After the discovery of the second client, the message flow is again the same as in the first scenario.

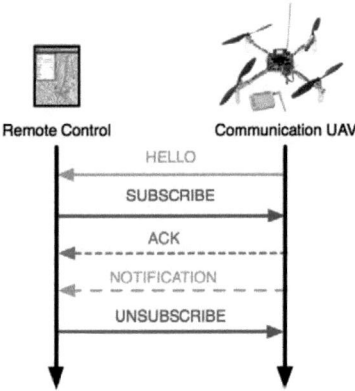

Figure 7.10: Message flow for subscribing to notification service.

In order to provide monitoring functionality of all UAVs in UAVNet, a remote control client can subscribe to a notification service on each UAV (see Figure 7.10). After discovering the UAV, a remote control client sends a *SUBSCRIBE* message to the discovered UAV. The UAV then acknowledges the request and adds the remote control client to the subscriber list. Henceforth, the UAV sends unicast messages containing state information to all subscribers.

7.5 Remote Control Client

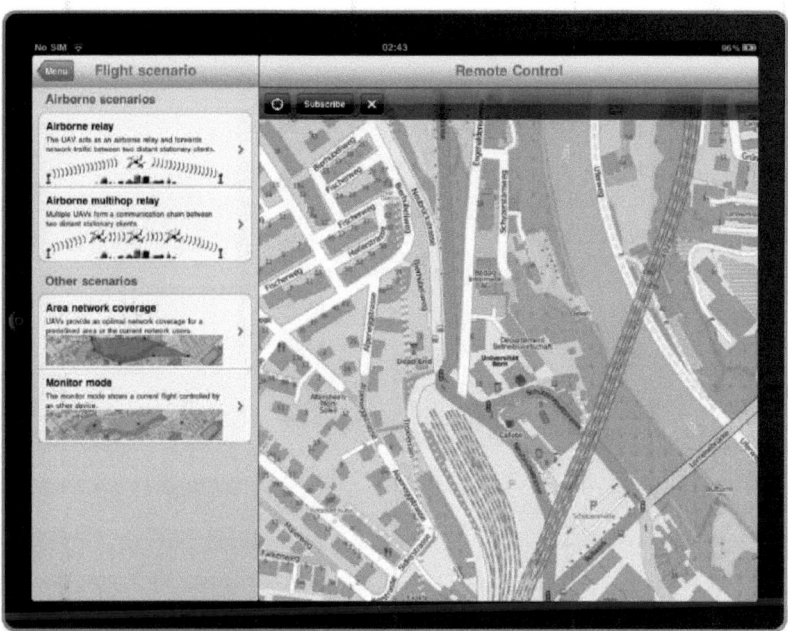

Figure 7.11: Remote Control Application (iPad): Selection of network deployment scenario.

In order to remotely control the UAVNet prototype, we developed a control client as an iPhone/iPad application, termed Remote Control App. It offers a convenient interface to use the UAVNet system. It is fully aware of the location and orientation of the iPhone/iPad. By using the GPS receiver of the device, the Remote Control App shows the current position of the user on an electronic map. This map is automatically oriented towards the magnetic North by employing the electronic compass of the device. Moreover, the Remote Control App adjusts its graphical user interface automatically to the device orientation (horizontal or vertical).

7.5. REMOTE CONTROL CLIENT

The representation of the electronic map is based on the route-me library [81] and can work with maps from different sources, e.g., the OpenStreetMap project [138]. Figure 7.11 shows the map displayed in landscape orientation on an iPad.

Figure 7.12: Remote Control Application (iPhone): setting the scenario, confirmation, and monitoring of flying UAVNet.

The control options for selecting the deployment scenario are shown on the left side of the electronic map (see Figure 7.11). The deployment options are airborne relay, multi-hop airborne relay, area coverage (as a support for future extensions of UAVNet), and monitoring. After selecting the appropriate scenario including search and positioning mode, the user sets up the scenario, selects the UAVs to be used, and starts the deployment process. Figure 7.12 depicts this process of the Remote Control App on an iPhone.

After the start of the deployment, the Remote Control App can monitor the entire UAVNet as it subscribes to the notification service on all UAVs. According to the received notification messages, the current state of the UAVs is graphically represented on the electronic map. Figure 7.13 illustrates the indication markers related to the UAVs characteristics: a blue arrow representing the current speed (length) and flight direction of the UAV, a red dot showing the exact GPS position, and the current altitude relative to the start location.

7.6. EVALUATION

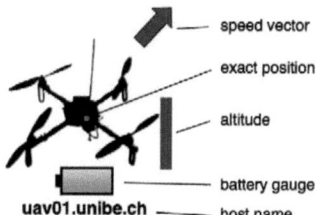

Figure 7.13: GUI-Marker representing the current state of a UAV.

7.6 Evaluation

In order to evaluate the performance of UAVNet, RTT and TCP/UDP throughput measurements have been performed. RTT is evaluated by using the standard *ping* tool from the *iputils* package [112]. Our evaluations are based on 1'000 *ping* measurements with a payload size of 56 bytes and a measurement interval of 0.1 seconds. The TCP and UDP throughput have been measured with the *netperf* measurement tool [102] using the *TCP_STREAM* and *UDP_STREAM* tests.

7.6.1 Determination of Optimal Signal Strength Thresholds

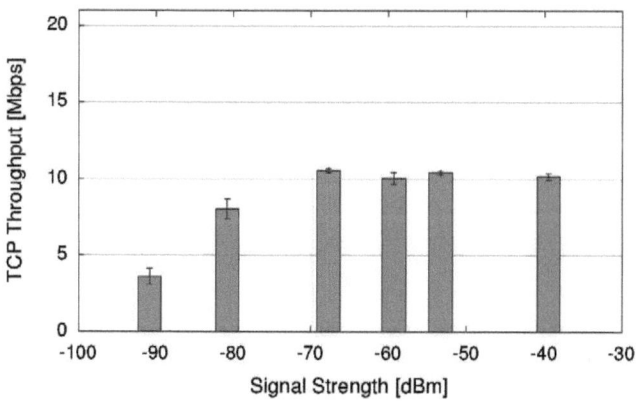

Figure 7.14: TCP throughput between two nodes depending on signal strength.

First, in order to find optimal locations (positions) of the UAVs, we need to

7.6. EVALUATION

establish reasonable thresholds for the signal strength. We, therefore, measured the TCP and UDP throughput and the RTT between two nodes while the distance between them was gradually decreased. The OM1P mesh nodes have an IEEE 802.11b/g radio. At the beginning of each measurement, the signal strength was determined by the Linux standard tool *iw*. Figure 7.14 depicts the average TCP throughput values and the corresponding standard deviations (whiskers) of *netperf* TCP_STREAM tests. For received signal strengths higher than -70 dBm, a TCP throughput of about 10 Mbps has been reached. For lower signal strength values, the TCP throughput decreases significantly as more packet retransmissions are required to recover lost packets. The maximum TCP throughput in Figure 7.14 is lower than measured in OViS (22 Mbps) as an OM1P is less powerful than an ALIX node.

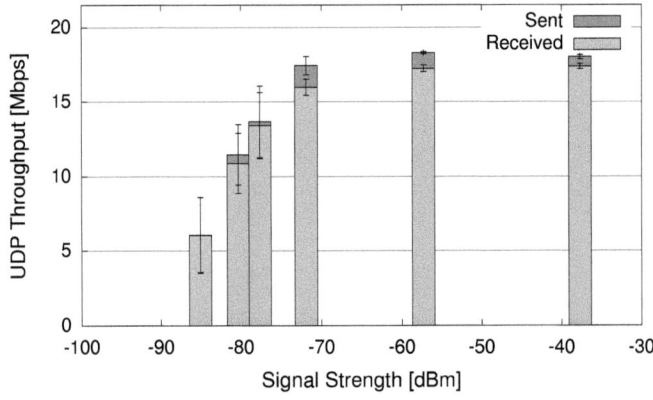

Figure 7.15: UDP throughput between two nodes depending on signal strength.

The corresponding results for the UDP throughput, given the same scenario, are shown in Figure 7.15. For a received signal strength higher than -70 dBm, an average UDP throughput of 16 - 17.4 Mbps is reached, which is significantly higher than for TCP as expected. Similar to the TCP throughput measurements, the average UDP throughput values also significantly decrease for received signal strength lower than -70 dBm. As UDP has no flow and congestion control, the reported send rate is higher than the actual receive rate in the *UDP_STREAM* test of *netperf*. Thus, some packets are lost.

Together with the TCP and UDP throughput, the RTT was also measured. Figure 7.16 depicts the average RTT values and the standard deviations. For signal strengths lower than -70 dBm, the values increase as the automatic data rate control

7.6. EVALUATION

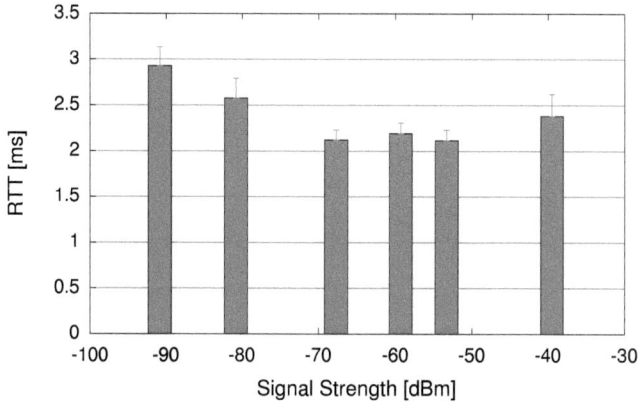

Figure 7.16: RTT between two stationary nodes depending on signal strength.

switches to lower data rates.

Based on the performed measurements and our experiences with OViS, we selected a signal strength threshold of -60 ± 10 dBm to be used for the positioning of the UAVs.

7.6.2 Multi-Hop Performance

Second, we used the determined signal strength threshold to evaluate the multi-hop performance of the proposed flying WMN. We used IEEE 802.11s with the airtime metric on the OM1P nodes. Four stationary nodes were placed outdoors in a chain topology using the determined optimal signal strength threshold of -60 ± 10 dBm. The TCP and UDP throughput were measured between the first node (node01) and the other three remaining nodes (node02, node03, node04). The results in Figure 7.17 represent the average values and the standard deviations. As our UAVNet prototype uses only one radio and single channel communication, an expected throughput decrease, due to the channel sharing, is observed. For example, the throughput significantly decreases in a two-hop scenario as node02 now has to forward the packets to node03 using the same radio as for the communication with node01. Interestingly, there is only a small decrease when communicating over three hops compared to two hops. The communication on the last link between node03 and node04 is almost not affected by the communication on the first link due to the positioning of the nodes.

7.7. CONCLUSIONS

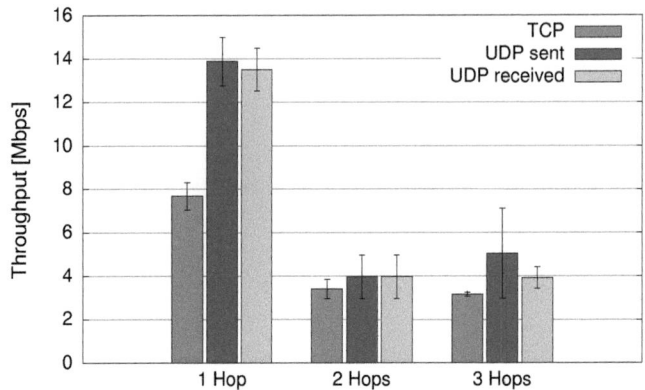

Figure 7.17: TCP and UDP throughput over multiple hops.

7.6.3 Effect of Too Far Away Nodes

To show the effect of too far away nodes, we placed two nodes (node01, node03) at the maximum distance defined by their transmission ranges, i.e., at a signal strength of about -100 dBm. The nodes were deployed outdoors and remained at the same position throughout the experiment. We then measured the TCP and UDP throughput between these two nodes as well as with an intermediate node (node02) in between. Figure 7.18 shows a significantly increased throughput in case of an additional intermediate node. Whereas the direct connection provides a TCP and UDP throughput of less than 0.3 Mbps, the 2-hop connection provides TCP throughput of 2.7 Mbps and a UDP throughput of 3.7 Mbps, which is enough for emergency communication, including video conferencing. The throughput values match the ones of Figure 7.17.

7.7 Conclusions

In this chapter, we proposed a concept for an autonomous deployment of a temporary flying IEEE 802.11s WMN for emergency and disaster recovery scenarios. The concept is based on WMN nodes that are carried by small quadrocopter UAVs. The on-board WMN node is connected to the electronic autopilot of the UAV in order to enable the autonomous network deployment. Thus, the WMN node can indirectly control the flight of the UAV by adding/removing navigation points. The network

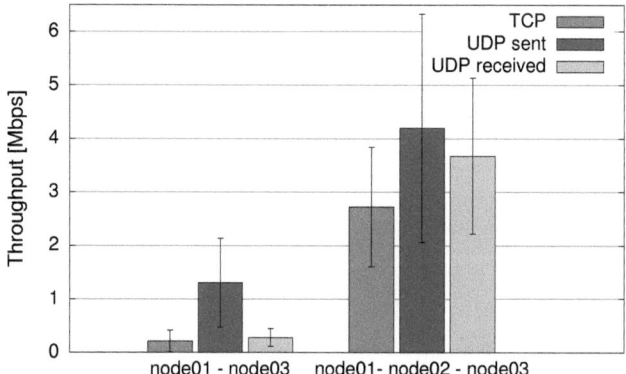

Figure 7.18: TCP and UDP throughput between two distant nodes.

deployment is coordinated among the flying WMN nodes and the remote control client over the IEEE 802.11s WMN.

We have proven the feasibility of a flying network by a prototype implementation based on small quadrocopter UAVs and OpenMesh OM1P mesh nodes, communicating over IEEE 802.11s. This prototype autonomously interconnects two distant clients via one or multiple airborne relays. The network deployment is initialised depending on the desired scenario using a user-friendly remote control application running on iPhone/iPad devices. The remote control application further provides monitoring of the flying WMN.

All further development of the UAVNet prototype can be excellently supported by VirtualMesh, our wireless driver-enabled network emulation framework (see Chapter 4). VirtualMesh offers testing of real UAVNet implementations without the risk of crashing costly UAVs and in a larger scale than normally possible in a real testbed. In order to fully support the testing of UAVNet, VirtualMesh has to be extended to support UAVs in the simulation model and propagating their positions.

There are several possible extensions for our prototype of UAVNet. First, a replacement and recharging strategy would keep the system working without assistance for several hours or days. UAVs with low battery capacity should automatically leave the formation and fly to a recharge station. The formation adapts automatically to ensure network connectivity. After recharging, the UAVs can be reintegrated in UAVNet. Second, a strategy for maximising network coverage of a specified area can be introduced. The UAVs should autonomously position them-

7.7. CONCLUSIONS

selves over a user-defined area such that to maximise the network coverage, given the number of available UAVs. The UAV swarm can further adapt to the current communication needs by placing more UAVs in areas with more clients than in sparsely populated areas.

In addition, UAVNet provides a basic prototype for our research project "Opportunistic Routing for Highly Mobile Ad-hoc Networks" (ORMAN) and helps in developing appropriate topology control algorithms and opportunistic multi-channel routing protocols for highly mobile networks. Topology control should guarantee that the UAV swarm always forms an interconnected communication network. It may be based on the concept of virtual springs taking the inverse RSSI values as spring forces to adjust the swarm constantly to the new environment. New opportunistic routing protocols should take advantage of all available data from the UAV, e.g., position, speed, flight direction, and altitude.

Chapter 8
Conclusions and Outlook

Wireless mesh networks are a key technology to achieve ubiquitous broadband network access for various application scenarios. WMNs and the services running on top of them undergo a common life cycle covering development, testing, deployment, and operation. In order that WMNs can be pervasively used, the phases in their life cycle requires appropriate tools, architectures, best practises, and tested equipment. In this thesis, we addressed the following challenges in the life cycle of a WMN:

- Heterogeneous embedded hardware platforms for WMN nodes with limited capabilities have to be supported in a WMN. We addressed the challenge of heterogeneity by a comprehensible cross-compilation build system for an embedded Linux system tailored for WMN nodes. It provides the same software for different hardware platforms. By storing the software in a compressed read-only image, even low cost nodes with only 8 MB of permanent storage can provide similar software functionality as more powerful nodes.

- Guaranteeing remote accessibility to all network nodes in the presence of faulty configuration and software updates to prevent costly on-site repairs is a major challenge during network operation. Our solution employs a decentralised distribution mechanism for updates and self-healing mechanisms.

- Prototype implementations have to be adequately, iteratively and flexibly tested under various conditions. As this is not possible in a real testbed with a limited scale, or only at high costs, we introduced the concept of wireless device driver enabled network emulation. It combines the flexibility and scalability of network simulation and testing the real prototype. Our solution provides a high integration through a virtual device driver. This virtual driver replaces the standard wireless device driver and redirects the network traffic and all device parameters to a simulation model, which emulates the wireless medium and provides support for node mobility.

- Various challenges concerning an outdoor deployment, e.g., equipment, software or mechanical problems, and environmental conditions, were addressed by providing best practises and deployment experiences for future deployments.

- The deployment of a WMN is not trivial and requires expert knowledge. In order to enable non-experts to rapidly deploy a temporary WMN, we introduced a deployment procedure using an electronic guide that instructs the user through the deployment.

- Some application scenarios, such as disaster recovery, require immediate availability of a broadband communication infrastructure. As a guided WMN deployment procedure may too time-consuming, the network has to be automatically established without manual interaction. We implemented a prototype of a flying WMN using small quadrocopter UAVs carrying the mesh nodes. The prototype provides an autonomous deployment of a WMN between two remote clients.

To provide solutions to all these challenges, we developed a comprehensive WMN framework consisting of the following components:

- **ADAM Build System:** A modular cross-compilation build system for an embedded Linux distribution, tailored for nodes of a WMN.

- **ADAM Management:** A management architecture for safe and fault-tolerant configuration and software updates within a WMN.

- **VirtualMesh:** An inexpensive testing infrastructure based on network emulation and a virtual wireless interface driver, offering a transparent replacement of the real driver, i.e., dynamic propagation of wireless settings to the network emulation.

- **CTI-Mesh:** Documented experiences, best practises, a tested deployment process, tested equipment, and tested software obtained during the deployment of an outdoor WMN for environmental monitoring.

- **OViS:** A battery powered WMN supporting semi-automated (guided) network deployment by non-expert users.

- **UAVNet:** An autonomous deployment concept and architecture for a WMN using flying robots carrying the mesh nodes.

8.1 Summary

Our first contribution includes the ADAM management architecture, which avoids costly on-site repairs and reconfigurations in WMNs due to misconfiguration, corrupt software updates, or unavailability of nodes during updates (see Chapter 3). The node's accessibility is guaranteed by using a decentralised distribution mechanism for software and configuration updates, self-healing mechanisms, and a safe software update procedure. The management in ADAM is performed completely in-band, i.e., it does not require any additional network co-located for management. Moreover, the decentralised distribution of configuration and software updates works completely independent of routing mechanisms. Updates are efficiently distributed by splitting the node's firmware into node specific and type specific parts. Isolated nodes, e.g., due to misconfiguration, automatically re-join the WMN using recovery mechanisms. Faulty software updates are automatically recovered by the mesh node using a boot loader mechanism. In contrast to existing management solutions, ADAM improves the availability of the network independently of configuration errors and faulty software updates and works completely in-band. ADAM significantly reduces the number of costly on-site management actions and the overall network operation costs.

In addition to ADAM management, we developed a cross-compilation build system for an embedded Linux distribution for several mesh node types (see Chapter 3). The ADAM build system is modular, user-friendly, easy to understand, and extendable. It perfectly supports the developer in compiling software for heterogeneous WMN hardware platforms.

VirtualMesh provides adequate, iterative and flexible testing of prototype implementations under various conditions (see Chapter 4). It significantly simplifies the testing process by combining network simulation, providing a controlled environment and scalability, and prototype testing in a real testbed. VirtualMesh is based on traffic interception and redirection by a virtual wireless driver and emulating the wireless medium by a network simulator. In contrast to other emulation approaches, VirtualMesh provides a high integration of the virtual driver and the network emulation model. The virtual network driver is completely transparent for the operating system, i.e., all configuration changes are directly propagated to the simulation model and influence the network emulation. This is currently a unique feature, which can only be found in expensive hardware emulation. Our evaluations proved that VirtualMesh is a valuable testing architecture, which can be used prior to evaluations in a real test-bed or the final deployment in the productive network.

In CTI-Mesh, we have proven the feasibility and applicability of a WMN for interconnecting remote sensors to a fibre based network backbone over a distance of more than 20 km (see Chapter 5). The solar-powered WMN with directional links provided the requested robust network service for transmitting weather data.

8.2. OUTLOOK

Valuable contributions to the research and development community are the documentation of extensive deployment experiences, best practises for deployment, the tested deployment process, as well as the tested equipment and software. They represent a valuable starting point for any future WMN outdoor deployments. This knowledge helps in preventing common problems and pitfalls, resulting in significant time savings.

In OViS, we addressed the deployment of a temporary WMN by non-experts by using a guided deployment process (see Chapter 6). The temporary WMN includes battery-powered nodes that are deployed by non-expert users following the easily understandable instructions of a deployment wizard application running on a handheld or smart phone device. The network is then automatically configured to use multi-channel communication. We have proven the feasibility of the semi-automated, i.e., guided, network deployment by non-expert users. An OViS network can be used in various situations, where there is a need for a temporary broadband network to support on-site operation, e.g., installation on construction sites, in mines, tunnels etc.

UAVNet provides an autonomous deployment concept and architecture for a WMN using flying robots carrying the mesh nodes for emergency and disaster recovery scenarios (see Chapter 7). We have proven the functionality of such a flying network by a prototype implementation based on small quadrocopter UAVs and OpenMesh OM1P mesh nodes, communicating over IEEE 802.11s. UAVNet autonomically deployed a network that interconnected two distant end systems. The UAVNet prototype offers starting network deployment according to different scenarios using a user-friendly remote control application running on iPhone/iPad devices. The application monitors flying mesh nodes. UAVNet provides a basic prototype for our research project "Opportunistic Routing for Highly Mobile Ad-hoc Networks" (ORMAN).

In summary, we developed tools to support the development, testing, deployment and operation of WMNs for various application scenarios, e.g., environmental monitoring, construction sites, and disaster recovery. Our contributions bring WMNs closer to be pervasively deployed in various application scenarios.

8.2 Outlook

There are several possible directions for future research and development starting from our work. ADAM Linux and management can be extended to support the configuration of more routing protocols, to support more software packages and to include multi-channel protocols for communication. In order to simplify the addition of new software, the development of an automatic converter of OpenWrt software packages could be envisioned. The ADAM build system can be extended to provide

8.2. OUTLOOK

a dependency calculator for software packages. ADAM can be further extended to support additional platforms. Currently, ARM Cortex-A8 based Gumstix Overo computer-on-modules are being added as a new target platform for the project "Location Based Analyzer".

In order to increase the applicability of VirtualMesh to additional testing scenarios, VirtualMesh could be extended to synchronised network emulation, bi-directional propagation of wireless parameters, and the extension to other network technologies.

Although VirtualMesh proved high scalability when the simulation model is run on a powerful machine and the communication runs over a dedicated high performance network, complex scenarios or sophisticated radio models may still lead to overload situations, where the simulation model cannot keep pace with the injected traffic. A solution is to integrate the concept of synchronised network emulation [200] into VirtualMesh. In this approach, a central synchronising component, connected to the simulation model, controls the time flow for the involved virtualised nodes. This offers testing in larger topologies using complex radio propagation models on commodity hardware, keeping the benefits of testing the real prototype implementations.

Support for scanning for available networks, passive scanning of a wireless interface in the promiscuous mode (channel sniffing) and retrieval of SNR values requires propagation of the wireless parameters from the simulation model back to the nodes. VirtualMesh currently only supports the propagation in the opposite direction. An extension to support bi-directional exchange of the wireless parameters would extend the applicability of VirtualMesh to further scenarios.

The current implementation of the VirtualMesh supports IEEE 802.11b/g networks. As the concept is quite flexible, it can be easily extended to support other network technologies, e.g., WiMAX, Bluetooth or LTE. The support of additional technologies requires additional simulation models and adapted virtual drivers that match the APIs of the real drivers of the corresponding technologies. Another open issue is the support for the new Netlink-based wireless configuration interface of the Linux kernel for IEEE 802.11a/b/g/h/n devices.

CTI-Mesh showed that the deployment of an outdoor WMN with directional antennas is a time-consuming process, which could be simplified by an appropriate deployment application on a smart phone. For example, the correct alignment of the antennas is a cumbersome task. After a first alignment with a compass, fine-tuning is performed by stepwise turning the antenna in one direction while measuring the received signal strength at the outstation. Using a notebook for this task is not very practical due to its size / weight. The antenna alignment process could be supported and simplified by a smart phone application. The application could provide easy understandable visual and acoustic instructions as in OViS. To receive signal strength measurements, it either automatically connects to the outstation or

8.2. OUTLOOK

triggers a probing request from the station currently deployed. With the help of such deployment wizards, WMNs with directional antennas could be deployed by non-expert users.

Instead of assigning the channels statically during the deployment process, OViS could make use of multi-channel MAC protocols to adapt the channel assignment dynamically during the network lifetime. The concept of OViS could be enhanced to provide network coverage of an area instead of only bridging a distance.

The prototype of UAVNet offers various possibilities for research. Replacement and recharging strategies for the UAV swarm are necessary to keep the network working for several hours. New topology control algorithms are required to adapt the network to changing communication needs or optimally covering an area. Algorithms for area coverage could be based on virtual springs using measured signal strengths as spring forces. Furthermore, new opportunistic routing protocols could take advantage of all available data from the UAV, e.g., position, speed, flight direction, and altitude.

Chapter 9
Acronyms

ACI	Adjacent Channel Interference
ADAM	Administration and Deployment of Adhoc Mesh networks
AODV	Ad hoc On-demand Distance Vector Routing
ARP	Address Resolution Protocol
CAPWAP	Control And Provisioning of Wireless Access Points
CHAT	CHMA with Packet Train
CHMA	Channel Hopping Multiple Access
CLFS	Cross Linux From Scratch
CPU	Central Processing Unit
CTI	Swiss Commission for Technology and Innovation
DAMON	Distributed Ad-hoc Network Monitoring
DCA	Dynamic Channel Allocation
DFS	Dynamic Frequency Selection
DHCP	Dynamic Host Configuration Protocol
DNS	Domain Name System
DSDV	Destination-Sequenced Distance Vector Routing
DSL	Digital Subscriber Line

DSR	Dynamic Source Routing	
EIRP	Equivalent Isotropically Radiated Power	
ELF	Executable and Linkable Format	
ETT	Expected Transmission Time	
ETX	Expected Transmission Count	
FEC	Forward Error Correction	
FPGA	Field Programmable Array	
FTP	File Transfer Protocol	
GCC	GNU Compiler Collection	
GNU	GNU is Not Unix	
GPS	Global Positioning System	
HMCP	Hybrid Multichannel Protocol	
HNA	Host and Network Association	
HRMA	Hop-Reservation Multiple Access	
HTTP	Hypertext Transfer Protocol	
HWMP	Hybrid Wireless Mesh Protocol	
IEEE	Institute of Electrical and Electronics Engineers	
IP	Internet Protocol	
IPv4	Internet Protocol version 4	
IPv6	Internet Protocol version 6	
JiST	Java in Simulation Time	
LAN	Local Area Network	
LFS	Linux From Scratch	
LQSR	Link Quality Source Routing	

MANET	Mobile Ad-hoc Network
MAP	Multichannel Access Protocol / Mesh Access Point (IEEE 802.11s)
MCL	Mesh Connectivity Layer
MMAC	Multichannel MAC Protocol
MPR	Multi-Point-Relays
MTU	Maximum Transfer Unit
NAT	Network Address Translator
NFS	Network File System
OE	OpenEmbedded
OFCOM	Swiss Federal Office of Communication
OViS	On-site Video System
PC	Personal Computer
QoS	Quality of Service
RAM	Random Access Memory
ROMER	Resilient Opportunistic Mesh Routing
RREP	Route Reply
RREQ	Route Request
RSSI	Received Signal Strength Indicator
RTT	Round Trip Time
SMR	Split Multi-Path Routing
SNMP	Simple Network Management Protocol
SNR	Signal-to-Noise Ratio
SSCH	Slotted Seeded Hopping
TBRPF	Topology Broadcast based on Reverse-Path Forwarding

TC	Topology Control
TCP	Transport Control Protocol
TFA	Technology for All
TPC	Transmit Power Control
TTL	Time to Live
UAV	Unmanned Arial Vehicle
UDP	User Datagram Protocol
USB	Universal Serial Bus
WCETT	Weighted Expected Transmission Count
WMN	Wireless Mesh Network
WRAP	Wireless Router Application Platform
WSN	Wireless Sensor Network

Bibliography

[1] D. Aguayo, J. Bicket, S. Biswas, D. S. J. D. Couto, and R. Morris, "MIT Roofnet Implementation," http://pdos.lcs.mit.edu/roofnet/design/, August 2003.

[2] D. Aguayo, J. Bicket, S. Biswas, G. Judd, and R. Morris, "Link-Level Measurements from an 802.11b Mesh Network," in *International Conferences on Broadband Networks (BroadNets)*, San José, CA, USA, October 25-29 2004.

[3] I. F. Akyildiz and X. Wang, "A Survey on Wireless Mesh Networks," *Communications Magazine, IEEE*, vol. 43, no. 9, pp. 23–30, 2005.

[4] I. F. Akyildiz, X. Wang, and W. Wang, "Wireless Mesh Networks: a Survey," *Computer Networks Journal (Elsevier)*, vol. 47, no. 4, pp. 445–487, 15 March 2005.

[5] E. Andersen, "μClibc: A C Library for Embedded Linux," http://www.uclibc.org, 2011.

[6] E. Andersen, R. Landley, B. Reutner-Fischer, D. Vlasenko, and various developers, "BusyBox," http://www.busybox.net/, 2011.

[7] V. Angelakis, S. Papadakis, N. Kossifidis, V. A. Siris, and A. Traganiti, "The Effect of Using Directional Antennas on Adjacent Channel Interference in 802.11a: Modeling and Experience With an Outdoors Testbed," in *4th International Workshop on Wireless Network Measurements (WiNMee 2008)*, Berlin, Germany, March 2008.

[8] V. Angelakis, M. Genetzakis, N. Kossifidis, K. Mathioudakis, M. Ntelakis, S. Papadakis, N. Petroulakis, and V. A. Siris, "Heraklion MESH: An Experimental Metropolitan Multi-Radio Mesh Network," in *2nd ACM International Workshop on Wireless Network Testbeds, Experimental evaluation and CHaracterization (WINTECH 2007)*, Montreal, QC, Canada, September 10 2007.

BIBLIOGRAPHY

[9] Apple Inc., "iOS Operating System," http://developer.apple.com/devcenter/ios, April 2011.

[10] V. Aseeja and R. Zheng, "MeshMan: A Management Framework for Wireless Mesh Networks," in *IFIP/IEEE International Symposium on Integrated Network Management (IM '09)*, Long Island, New York, USA, June 1-5 2009, pp. 226–233.

[11] P. Bahl, R. Chandra, and J. Dunagan, "SSCH: Slotted Seeded Channel Hopping for Capacity Improvement in IEEE 802.11 Ad-Hoc Wireless Networks," in *10th Annual International Conference on Mobile Computing and Networking (MobiCom '04)*. Philadelphia, PA, USA: ACM, September 26 - October 1 2004, pp. 216–230.

[12] M. Bahr, "Update on the Hybrid Wireless Mesh Protocol of IEEE 802.11s," in *4th IEEE International Conference on Mobile Adhoc and Sensor Systems (MASS 2007)*, Pisa, Italy, Oct. 2007, pp. 1–6.

[13] A. Baiocchi, A. Todini, and A. Valletta, "Why a Multichannel Protocol can Boost IEEE 802.11 Performance," in *7th ACM international symposium on Modeling, Analysis and Simulation of Wireless and Mobile Systems (MSWiM '04)*, Venice, Italy, October 4-6 2004, pp. 143–148.

[14] M. Baker, G. Rozema, and various developers. (2011) OpenWrt: a Linux Distribution for Embedded Devices. http://openwrt.org/.

[15] D. Balsiger, "Administration and Deployment of Wireless Mesh Networks," Master's thesis, University of Bern, Bern, Switzerland, April 2009.

[16] D. Balsiger and M. Lustenberger, "Secure Remote Management and Software Distribution for Wireless Mesh Networks," Bachelor's thesis, University of Bern, Bern, Switzerland, September 2007.

[17] P. Barham, B. Dragovic, K. Fraser, S. Hand, and T. Harris, "Xen and the Art of Virtualization," in *9th ACM Symposium on Operating Systems Principles (SOSP '03)*. Bolton Landing, NY, USA: ACM, October 19 - 22 2003, pp. 164–177.

[18] R. Barr, Z. J. Haas, and R. van Renesse, "JiST: Embedding Simulation Time into a Virtual Machine," in *EUROSIM Congress on Modelling and Simulation*, Noisy-le-Grand, Paris, France, September 6-10 2004.

[19] G. Beekmans, M. Burgess, J. Gifford, J. Huntwork, Archaic, K. Moffat, M. C. Esparcia, R. Oliver, and N. Coulson, "Linux From Scratch (LFS)," http://www.linuxfromscratch.org, 2011.

BIBLIOGRAPHY

[20] E. M. Belding-Royer, K. C. Almeroth, H. Lundgren, K. Ramachandran, A. Jardosh, M. Benny, and A. Hewatt, "UCSB MeshNet," http://moment.cs.ucsb.edu/meshnet/, April 2011.

[21] J. Berg, "iw: CLI Configuration Utility for Wireless Devices," http://wireless.kernel.org/en/users/Documentation/iw, 2010.

[22] R. Beuran, L. T. Nguyen, T. Miyachi, J. Nakata, K.-I. Chinen, Y. Tan, and Y. Shinoda, "QOMB: A Wireless Network Emulation Testbed," in *Global Telecommunications Conference (GLOBECOM 2009)*, Honolulu, Hawaii, USA, November 30 - December 4 2009, pp. 1–6.

[23] R. Beuran, J. Nakata, T. Okada, L. T. Nguyen, Y. Tan, and Y. Shinoda, "A Multi-Purpose Wireless Network Emulator: QOMET," in *International Conference on Advanced Information Networking and Applications Workshops*, vol. 0. Gino-Wan, Okinawa, Japan: IEEE Computer Society, March 25-28 2008, pp. 223–228.

[24] R. Beuran, L. T. Nguyen, K. T. Latt, J. Nakata, and Y. Shinoda, "Qomet: A versatile wlan emulator," *Advanced Information Networking and Applications, International Conference on*, vol. 0, pp. 348–353, 2007.

[25] J. C. Bicket, D. Aguayo, S. Biswas, and R. Morris, "Architecture and Evaluation of an Unplanned 802.11b Mesh Network," in *11th Annual International Conference on Mobile Computing and Networking (MOBICOM 2005)*, Cologne, Germany, August 28 - September 2 2005, pp. 31–42.

[26] J. Blackford, H. Kirksey, and W. Lupton, "TR-069 Amendment 3 - CPE WAN Management Protocol," The Broadband Forum, Fremont, California, USA, Tech. Rep. 1, November 2010.

[27] R. Bless and M. Doll, "Integration of the FreeBSD TCP/IP-stack into the Discrete Event Simulator OMNet++," in *36th Winter Simulation Conference (WSC'04)*, Washington, D.C., USA, December 5-8 2004, pp. 1556–1561.

[28] B. Blywis, M. Güneş, F. Juraschek, and J. Schiller, "Trends, Advances, and Challenges in Testbed-based Wireless Mesh Network Research," *ACM/Springer Mobile Networks and Applications (MONET)*, February 2010, Special Issue on Advances in Wireless Testbeds and Research Infrastructures.

[29] K. Borries, G. Judd, D. Stancil, and P. Steenkiste, "FPGA-Based Channel Simulator for a Wireless Network Emulator," in *IEEE 67th Vehicular Technology Conference (VTC2009-Spring)*, Barcelona, Catalunya, Spain, April 2009.

BIBLIOGRAPHY

[30] T. Braun, G. Coulson, and T. Staub, "Towards Virtual Mobility Support in a Federated Testbed for Wireless Sensor Networks," in *6th Workshop on Wireless and Mobile Ad-Hoc Networks (WMAN 2011)*. Kiel, Germany: Electronic Communications of EASST, March 10 2011.

[31] T. Braun, M. Diaz, J. E. Gabeiras, and T. Staub, *End-to-End Quality of Service Over Heterogeneous Networks*. Springer, August 2008.

[32] T. X. Brown, B. Argrow, C. Dixon, S. Doshi, R. george Thekkekunnel, and D. Henkel, "Ad hoc UAV Ground Network (AUGNet)," in *AIAA 3rd "Unmanned Unlimited" Technical Conference*, Chicago, IL, USA, September 20-23 2004.

[33] R. Bruno, M. Conti, and E. Gregori, "Mesh Networks: Commodity Multhop Ad Hoc Networks," *IEEE Communications Magazine*, vol. 43, no. 3, pp. 123–131, March 2005.

[34] M. Burgess, "A Tiny Overview of Cfengine: Convergent Maintenance Agent," in *1st International Workshop on Multi-Agent and Robotic Systems MARS/ICINCO*, Barcelona, Spain, September 2005.

[35] H. Buss and I. Busker, "Mikrokopter Platform," http://www.mikrokopter.de, April 2011.

[36] ——, "Mikrokopter Serial Protocol," http://mikrokopter.de/ucwiki/en/SerialProtocol, April 2011.

[37] C. Larson, M. Lauer, et al., "OpenEmbedded Project," http://www.openembedded.org, 2011.

[38] C. Larson, Phil Blundell, et al., "BitBake - a Generic Task Executor," http://developer.berlios.de/projects/bitbake, 2011.

[39] P. Calhoun, M. Montemurro, and D. Stanley, "Control and Provisioning of Wireless Access Points (CAPWAP) Protocol Binding for IEEE 802.11," RFC 5416 (Proposed Standard), Internet Engineering Task Force, Mar. 2009. [Online]. Available: http://www.ietf.org/rfc/rfc5416.txt

[40] ——, "Control And Provisioning of Wireless Access Points (CAPWAP) Protocol Specification," RFC 5415 (Proposed Standard), Internet Engineering Task Force, Mar. 2009. [Online]. Available: http://www.ietf.org/rfc/rfc5415.txt

BIBLIOGRAPHY

[41] J. Camp and E. Knightly, "The IEEE 802.11s Extended Service Set Mesh Networking Standard," *IEEE Communications Magazine*, vol. 46, no. 8, pp. 120–126, 2008.

[42] J. D. Camp, E. W. Knightly, and W. S. Reed, "Developing and Deploying Multihop Wireless Networks for Low-Income Communities," in *Digital Communities*, Napoly, Italy, June 2005.

[43] J. Case, R. Mundy, D. Partain, and B. Stewart, "Introduction and Applicability Statements for Internet-Standard Management Framework," RFC 3410 (Informational), Internet Engineering Task Force, Dec. 2002. [Online]. Available: http://www.ietf.org/rfc/rfc3410.txt

[44] J. Case, M. Fedor, M. Schoffstall, and J. Davin, "Simple Network Management Protocol (SNMP)," RFC 1157 (Historic), Internet Engineering Task Force, May 1990. [Online]. Available: http://www.ietf.org/rfc/rfc1157.txt

[45] M. C. Castro, P. Dely, A. J. Kassler, F. P. Delia, and S. Avallone, "OLSR and Net-X as a Framework for Channel Assignment Experiments - Poster Presentation," in *WiNTECH 09*, Beijing, China, September 21 2009.

[46] M. C. Castro, P. Dely, A. J. Kassler, and N. H. Vaidya, "QoS-Aware Channel Scheduling for Multi-Radio/Multi-Channel Wireless Mesh Networks," in *WiNTECH 09*, Beijing, China, September 21 2009.

[47] M. C. Castro, A. Kassler, and S. Avallone, "Measuring the Impact of ACI in Cognitive Multi-Radio Mesh Networks," in *IEEE 72nd Vehicular Technology Conference (VTC)*, 2010.

[48] J. Chen, S.-T. Sheu, and C.-A. Yang, "A New Multichannel Access Protocol for IEEE 802.11 Ad Hoc Wireless LANs," in *14th IEEE International Symposium on Personal, Indoor and Mobile Radio Communications (PIMRC 2003)*, vol. 3, Bejing, China, September 7-10 2003, pp. 2291–2296.

[49] C.-M. Cheng, P.-H. Hsiao, H. Kung, and D. Vlah, "Adjacent Channel Interference in Dual-radio 802.11a Nodes and Its Impact on Multi-hop Networking," in *IEEE GLOBECOM 2006*, San Francisco, CA, USA, 27 November - 1 December 2006.

[50] C. Chereddi, P. Kyasanur, and N. H. Vaidya, "Design and Implementation of a Multi-Channel Multi-Interface Network," in *2nd International Workshop on Multi-Hop Ad Hoc Networks: From Theory to Reality (REALMAN '06)*. Florence, Italy: ACM Press, May 26 2006, pp. 23–30.

BIBLIOGRAPHY

[51] ——, "Net-X: a Multichannel Multi-Interface Wireless Mesh Implementation," *SIGMOBILE Mob. Comput. Commun. Rev.*, vol. 11, no. 3, pp. 84–95, 2007.

[52] S. Cheshire, B. Aboba, and E. Guttman, "Dynamic Configuration of IPv4 Link-Local Addresses," RFC 3927 (Proposed Standard), Internet Engineering Task Force, May 2005. [Online]. Available: http://www.ietf.org/rfc/rfc3927.txt

[53] C. D. T. Clausen and P. Jacquet, "The Optimized Link State Routing Protocol version 2," IETF Draft RFC draft-ietf-manet-olsrv2-11, April 2010.

[54] T. Clausen and P. Jacquet, "Optimized Link State Routing Protocol (OLSR)," RFC 3626 (Experimental), Internet Engineering Task Force, Oct. 2003. [Online]. Available: http://www.ietf.org/rfc/rfc3626.txt

[55] G. Coulson, T. Braun, and T. Staub, "Adding Virtual Mobility to a Federated Testbed for Wireless Sensor Networks: a Proposal," Universität Bern, Institut für Informatik und angewandte Mathematik, Bern, Switzerland, Tech. Rep. IAM-10-004, August 2010.

[56] H. A. Council, "Interconnect Analysis: 10GigE and InfiniBand in High Performance Computing," http://www.hpcadvisorycouncil.com/pdf/IB_and_10GigE_in_HPC.pdf, 2009, White Paper.

[57] J. Crichigno, M.-Y. Wu, and W. Shu, "Protocols and Architectures for Channel Assignment in Wireless Mesh Networks," *Ad Hoc Networks*, vol. 6, pp. 1051–1077, September 2008.

[58] F. Damiani and P. Giannini, "Tkinter - Python's De-Facto Standard GUI (Graphical User Interface) Package," http://www.python.org/topics/tkinter, April 2011.

[59] K. Daniel, B. Dusza, A. Lewandowski, and C. Wietfeld, "AirShield: A System-of-Systems MUAV Remote Sensing Architecture for Disaster Response," in *IEEE International Systems Conference 2009 (SysCon)*. Vancouver, British Columbia, Canada: IEEE, March 2009, pp. 196 – 200.

[60] D. S. J. De Couto, D. Aguayo, J. Bicket, and R. Morris, "A High-Throughput Path Metric for Multi-Hop Wireless Routing," in *9th ACM International Conference on Mobile Computing and Networking (MobiCom '03)*, San Diego, California, September 14-19 2003.

[61] P. Dely, M. Castro, S. Soukhakian, A. Moldsvor, and A. Kassler, "Practical Considerations for Channel Assignment in Wireless Mesh Networks," in *IEEE*

BIBLIOGRAPHY

Broadband Wireless Access Workshop, held in conjunction with Globecom 2010, Miami, FL, USA, December 6-10 2010.

[62] P. Dely and A. Kassler, "KAUMesh Demo," in *9th Scandinavian Workshop on Wireless Ad-hoc Sensor Networks (Adhoc'09)*, Uppsala, Sweden, May 4-5 2009.

[63] P. Dely, A. Kassler, N. Bayer, H.-J. Einsiedler, and D. Sivchenko, "FUZPAG: A Fuzzy-Controlled Packet Aggregation Scheme for Wireless Mesh Networks," in *7th International Conference on Fuzzy Systems and Knowledge Discovery (FSKD'10)*, Yantai, China, August 10 - 12 2010.

[64] P. Dely, A. Kassler, N. Bayer, and D. Sivchenko, "An Experimental Comparison of Burst Packet Transmission Schemes in IEEE 802.11-based Wireless Mesh Networks," in *IEEE Global Telecommunications Conference GLOBECOM 2010*, Miami, FL, USA, December 6-10 2010.

[65] P. Dely, A. Kassler, and D. Sivchenko, "Theoretical and Experimental Analysis of the Channel Busy Fraction in IEEE 802.11," in *Future Network Mobile Summit*, Florence, Italy, June 16-18 2010.

[66] R. Draves, J. Padhye, and B. Zill, "Comparison of Routing Metrics for Static Multi-Hop Wireless Networks," in *Conference on Applications, Technologies, Architectures, and Protocols for Computer Communications SIGCOMM '04*. Portland, Oregon, USA: ACM Press, August 30 - September 3 2004, pp. 133–144.

[67] ——, "Routing in Multi-Radio, Multi-Hop Wireless Mesh Networks," in *10th Annual International Conference on Mobile Computing and Networking (MobiCom '04)*. Philadelphia, Pennsylvania, USA: ACM Press, September 26 - October 1 2004, pp. 114–128.

[68] R. Droms, "Dynamic Host Configuration Protocol," RFC 2131 (Draft Standard), Internet Engineering Task Force, Mar. 1997, updated by RFCs 3396, 4361, 5494. [Online]. Available: http://www.ietf.org/rfc/rfc2131.txt

[69] W. Dubowik, "Atheros: Fix ath5k Support on ar2315/2317," http://repo.or.cz/w/openwrt.git/commit/cf521fcca87ee5330d41200c3470ca78e6519eb3, April 2011.

[70] EN 60529:1991/A1, *Degrees of Protection Provided by Enclosures (IP Code) (IEC 60529:1989)*, European Committee for Standardization, 1991.

BIBLIOGRAPHY

[71] M. Engel, M. Smith, S. Hanemann, and B. Freisleben, "Wireless Ad-Hoc Network Emulation using Microkernel-Based Virtual Linux Systems," in *5th EUROSIM Congress on Modeling and Simulation*, Cite Descartes, Marne la Vallee, France, September 6-10 2004, pp. 198–203.

[72] ETSI, *Broadband Radio Access Networks (BRAN); 5 GHz High Performance RLAN; Harmonized EN Covering Essential Requirements of Article 3.2 of the R&TTE Directive (ETSI European Standard EN 301 893 V1.5.1)*, European Telecommunications Standards Institute, December 2008.

[73] R. Flickenger and et al, *Wireless Networking in the Developing World*, 2nd ed. wndw.net, 2007, 978-0-9778093-6-3.

[74] Freifunk Community, "Freifunk - Project for Free Wireless Networks and Frequencies (Open Spectrum)," http://freifunk.net, 2011.

[75] H. T. Friis, "A Note on a Simple Transmission Formula," in *Proceedings of the I.R.E. and Waves and Electrons*, vol. 34, no. 5, May 1946, pp. 254–256.

[76] E. Galstad and various developers, "Nagios," http://www.nagios.org/, April 2011.

[77] R. Gantenbein, "VirtualMesh: An Emulation Framework for Wireless Mesh Networks in OMNeT++," Master's thesis, University of Bern, Bern, Switzerland, June 2010.

[78] M. Gates, A. Warshavsky, A. Tirumala, J. Ferguson, J. Dugan, and various developers, "NLANR/DAST : iperf - The TCP/UDP Bandwidth Measurement Tool," http://iperf.sourceforge.net/, 2009.

[79] GCC Steering Committee, "GNU Compiler Collection (GCC)," http://gcc.gnu.org, 2011.

[80] A. Gecyasar, "Ad-Hoc Multipath Routing Protokolle," Bachelor's thesis, University of Bern, Bern, Switzerland, November 2006.

[81] J. Gentle, M. Tyson, V. Vyskocil, and various developers, "Route Me - Open Source iPhone-Native Slippy Map," https://github.com/route-me, April 2011.

[82] G. Giacobbi, "The GNU Netcat Project," http://netcat.sourceforge.net/, April 2011.

[83] L. Girod, T. Stathopoulos, N. Ramanathan, J. Elson, D. Estrin, E. Osterweil, and T. Schoellhammer, "A System for Simulation, Emulation, and Deployment of Heterogeneous Sensor Networks," in *2nd International Conference on*

BIBLIOGRAPHY

Embedded Networked Sensor Systems (SenSys '04). Baltimore, Maryland, USA: ACM Press, November 3-5 2004, pp. 201–213.

[84] Google Inc., "Android," http://www.android.com, April 2011.

[85] M. Güneş, F. Juraschek, B. Blywis, Q. Mushtaq, and J. Schiller, "A Testbed for Next Generation Wireless Networks Research," *Special Issue PIK on Mobile Ad-hoc Networks*, vol. 34, no. 4, 2009.

[86] O. Hahm, M. Güneş, and K. Schleiser, "DES-Testbed A Wireless Multi-Hop Network Testbed for Future Mobile Networks," in *5th GI/ITG KuVS Workshop on Future Internet*, Stuttgart, Germany, June 9 2010.

[87] A. Hänni, "iPad/iPhone App as a Frontend for Prototype of a Highly Adaptive and Mobile Communication Network using Unmanned Arial Vehicles (UAVs)," Bachelor's thesis, University of Bern, Bern, Switzerland, to be submitted.

[88] A. Hassan, "Simulations on Multipath Routing Based on Source Routing," Bachelor's thesis, University of Bern, Bern, Switzerland, August 2008.

[89] S. Hauert, S. Leven, J.-C. Zufferey, and D. Floreano, "Communication-based Leashing of Real Flying Robots," in *IEEE International Conference on Robotics and Automation (ICRA)*, 2010, pp. 15–20.

[90] S. Hauert, J.-C. Zufferey, and D. Floreano, "Evolved Swarming Without Positioning Information: An Application in Aerial Communication Relay," *Autonomous Robots*, vol. 26, no. 1, pp. 21–32, 2009.

[91] K. K. He, "Why and How to Use Netlink Socket," *Linux Journal*, 2005, http://www.linuxjournal.com/article/7356.

[92] G. Hiertz, D. Denteneer, S. Max, R. Taori, J. Cardona, L. Berlemann, and B. Walke, "IEEE 802.11s: The WLAN Mesh Standard," *Wireless Communications, IEEE*, vol. 17, no. 1, pp. 104 –111, February 2010.

[93] R. Hinden and B. Haberman, "Unique Local IPv6 Unicast Addresses," RFC 4193 (Proposed Standard), Internet Engineering Task Force, Oct. 2005. [Online]. Available: http://www.ietf.org/rfc/rfc4193.txt

[94] R. Hornig, "INET Framework for OMNeT++," http://inet.omnetpp.org/, 2010.

BIBLIOGRAPHY

[95] IEEE P802.11 Task Group S, "IEEE P802.11s™/ D5.0, draft amendment to standard IEEE 802.11™: Mesh Networking," IEEE, April 2010, work in progress.

[96] IEEE Standard Information Network, *IEEE 100 The Authoritative Dictionary of IEEE Standards Terms*, 7th ed. New York: IEEE: The Institute of Electrical and Electronics Engineers, 2000.

[97] S. Ivanov, A. Herms, and G. Lukas, "Experimental Validation of the ns-2 Wireless Model using Simulation, Emulation, and Real Network," in *4th Workshop on Mobile Ad-Hoc Networks (WMAN'07) in conjunction with the 15th ITG/GI - Fachtagung Kommunikation in Verteilten Systemen (KiVS'07)*. Bern, Switzerland: VDE Verlag, February 26 - March 2 2007, pp. 433–444.

[98] S. Jansen and A. McGregor, "Performance, Validation and Testing with the Network Simulation Cradle," in *14th IEEE International Symposium on Modeling, Analysis, and Simulation (MASCOTS '06)*. Monterey, California, USA: IEEE Computer Society, September 11-14 2006, pp. 355–362.

[99] D. Johnson, Y. Hu, and D. Maltz, "The Dynamic Source Routing Protocol (DSR) for Mobile Ad Hoc Networks for IPv4," RFC 4728 (Experimental), Internet Engineering Task Force, Feb. 2007. [Online]. Available: http://www.ietf.org/rfc/rfc4728.txt

[100] D. Johnson, K. Matthee, D. Sokoya, L. Mboweni, A. Makan, and H. Kotze, "Building a Rural Wireless Mesh Network - A Do-It-Yourself Guide to Planning and Building a Freifunk Based Mesh Network," http://wirelessafrica.meraka.org.za/wiki/images/f/fe/Building_a_Rural_Wireless_Mesh_Network_-_A_DIY_Guide_v0.7_65.pdf, October 30 2007.

[101] D. Johnson, T. Stack, R. Fish, D. M. Flickinger, L. Stoller, R. Ricci, and J. Lepreau, "Mobile Emulab: A Robotic Wireless and Sensor Network Testbed," in *25th IEEE International Conference on Computer Communications (INFOCOM 2006)*, Barcelona, Spain, April 23-29 2006.

[102] R. Jones, "netperf - Network Performance Benchmark," http://www.netperf.org/, April 2011.

[103] G. Judd and P. Steenkiste, "Repeatable and Realistic Wireless Experimentation through Physical Emulation," in *2nd Workshop on Hot Topics in Networks (Hot-Nets II)*, Boston, MA, USA, November 2003.

BIBLIOGRAPHY

[104] R. Karrer, A. Sabharwal, and E. Knightly, "Enabling Large-scale Wireless Broadband: The Case for TAPs," in *2nd Workshop on Hot Topics in Networks (Hot-Nets II)*, Cambridge, MA, USA, November 2003.

[105] A. Kassler, M. Castro, P. Dely, J. Karlsson, and A. Lavén, "KAUMesh," http://www.kau.se/en/kaumesh, 2011.

[106] J. Katz, "pyGrub," http://wiki.xensource.com/xenwiki/PyGrub, 2010.

[107] K. Kooi, T. Chick, M. Juszkiewicz, P. Sokolovsky, P. Balister, H. H. von Treskow, and B. Guillon, "The Ångström distribution," http://www.angstrom-distribution.org, April 2011.

[108] J. Krähenbühl, "Theory and Hands-on Exercises with Network Simulators for E-Learning on Distributed Systems," Master's thesis, University of Bern, Bern, Switzerland, September 2007.

[109] M. Krasnyansky and F. Thiel, *Universal TUN/TAP device driver*, 2002.

[110] T. Krop, M. Bredel, M. Hollick, and R. Steinmetz, "JiST/MobNet: Combined Simulation, Emulation, and Real-World Testbed for Ad Hoc Networks," in *WinTECH '07*. New York, NY, USA: ACM, 2007, pp. 27–34.

[111] G. S. Kulkarni, A. Nandan, M. Gerla, and M. B. Srivastava, "A Radio Aware Routing Protocol for Wireless Mesh Networks," UCLA Electrical Engineering, UCLA Computer Science, Los Angeles, CA, Tech. Rep. TR-UCLA-NESL-200503-12, March 2005.

[112] A. Kuznetsov and Y. Hideaki, "Linux iputils," http://www.linuxfoundation.org/collaborate/workgroups/networking/iputils, 2010.

[113] P. Kyasanur, J. So, C. Chereddi, and N. H. Vaidya, "Multichannel Mesh Networks: Challenges and Protocols," *IEEE Wireless Communications [see also IEEE Personal Communications]*, vol. 13, no. 2, pp. 30–36, April 2006.

[114] P. Kyasanur and N. H. Vaidya, "Routing and Interface Assignment in Multi-Channel Multi-Interface Wireless Networks," in *IEEE Wireless Communications and Networking Conference (WCNC 2005)*, vol. 4, New Orleans, Louisiana, USA, March 13 - 17 2005, pp. 2051–2056.

[115] P. Kyasanur, C. Chereddi, and N. H. Vaidya, "Net-X: System eXtensions for Supporting Multiple Channels, Multiple Interfaces, and Other Interface Capabilities," University of Illinois at Urbana-Champaign, Urbana, IL, USA, Tech. Rep., August 2006.

BIBLIOGRAPHY

[116] M. Lacage, M. Weigle, C. Dowell, G. Carneiro, G. Riley, T. Henderson, and J. Pelkey, "The Network Simulator ns-3," http://www.nsnam.org/, 2009.

[117] A. Lavén and A. Kassler, "Multi-channel anypath routing in wireless mesh networks," in *IEEE Globecom 2010 Workshop on Heterogeneous, Multi-hop Wireless and Mobile Networks (HeterWMN 2010)*, Miami, USA, December 6 2010.

[118] S.-J. Lee and M. Gerla, "Split Multipath Routing with Maximally Disjoint Paths in Ad Hoc Networks," in *IEEE International Conference on Communications (ICC)*, vol. 10, Helsinki, Finlandia, June 11-14 2001, pp. 3201–3205.

[119] H. Lundgren, E. Nordström, and C. Tschudin, "Coping with Communication Gray Zones in IEEE 802.11b Based Ad Hoc Networks," in *5th ACM International Workshop on Wireless Mobile Multimedia (WOWMOM '02)*. Atlanta, Georgia, USA: ACM, September 28 2002, pp. 49–55.

[120] J. Malinen, "Linux WPA/WPA2/IEEE 802.1X Supplicant," http://hostap.epitest.fi/wpa_supplicant/, 2010.

[121] D. Manzano, J.-C. Cano, C. Calafate, and P. Manzoni, "MAYA: A Tool For Wireless Mesh Networks Management," in *IEEE Internatonal Conference on Mobile Adhoc and Sensor Systems (MASS 2007)*, Pisa, Italy, October 8-11 2007, pp. 1–6.

[122] M. K. Marina and S. R. Das, "Ad hoc On-demand Multipath Distance Vector Routing," *ACM SIGMOBILE Mobile Computing and Communications Review*, vol. 6, no. 3, pp. 92–93, July 2002.

[123] R. McGrath, U. Drepper, and various developers, "GNU C Library (Glibc)," http://www.gnu.org/software/libc/, 2011.

[124] Meraki Inc., "Meraki 'free the net' Project in San Francisco," http://sf.meraki.com, 2007. [Online]. Available: http://meraki.com/about/freethenet/

[125] ——, "The Meraki Mini / Indoor Wireless Platform," http://meraki.com, 2007.

[126] Microsoft Research, "Mesh Connectivity Layer (MCL)," http://research.microsoft.com/en-us/projects/mesh/.

[127] P. Mochel, "The sysfs Filesystem," in *Proceedings of the 2005 Linux Symposium*, July 2005.

BIBLIOGRAPHY

[128] S. Morgenthaler, "Management Extensions for Wireless Mesh and Wireless Sensor Networks," Bachelor's thesis, University of Bern, Bern, Switzerland, March 2010.

[129] ——, "Prototype of a Highly Adaptive and Mobile Communication Network using Unmanned Aerial Vehicles (UAVs)," Master's thesis, University of Bern, Bern, Switzerland, to be submitted.

[130] C. Müller, "Implementation of a Multichannel Multiradio Prototype on Embedded Linux," Bachelor's thesis, University of Bern, Bern, Switzerland, May 2010.

[131] J. Nachtigall, A. Zubow, and J.-P. Redlich, "The Impact of Adjacent Channel Interference in Multi-Radio Systems using IEEE 802.11," in *Wireless Communications and Mobile Computing Conference (IWCMC '08)*, Crete Island, Greece, August 6-8 2008, pp. 874–881.

[132] OFCOM, *784.101.21 / RIR1010-04, 5470 - 5725 MHz, Wideband Data Transmission Systems*, 2nd ed., Federal Office of Communications (OFCOM), Switzerland, January 1st 2009.

[133] R. Ogier, F. Templin, and M. Lewis, "Topology Dissemination Based on Reverse-Path Forwarding (TBRPF)," RFC 3684 (Experimental), Internet Engineering Task Force, Feb. 2004. [Online]. Available: http://www.ietf.org/rfc/rfc3684.txt

[134] B. O'Hara, P. Calhoun, and J. Kempf, "Configuration and Provisioning for Wireless Access Points (CAPWAP) Problem Statement," RFC 3990 (Informational), Internet Engineering Task Force, Feb. 2005. [Online]. Available: http://www.ietf.org/rfc/rfc3990.txt

[135] A. Ollero, "Platform for Autonomous Self-Deploying and Operation of Wireless Sensor-Actuator Networks Cooperating with Aerial Objects," http://aware-project.net/, August 2009.

[136] Open-Mesh, "Open-Mesh OM1P," http://www.open-mesh.com/, 2011.

[137] Open80211s Consortium (Nortel, cozybit, one laptop per child, Google), "open80211s - A Reference Implementation of the Upcoming IEEE 802.11s Standard on Linux," http://open80211s.org/, April 2011.

[138] OpenStreetMap contributors, CC-BY-SA, "Map Data from OpenStreetMap," http://www.openstreetmap.org/, April 2011.

BIBLIOGRAPHY

[139] S. Ott, "Experimental Evaluation of Multi-Path Routing in a Wireless Mesh Network Inside a Building," Bachelor's thesis, University of Bern, Bern, Switzerland, February 2009.

[140] ——, "Automated Deployment of a Wireless Mesh Communication Infrastructure for an On-site Video-conferencing System (OViS)," Master's thesis, University of Bern, Bern, Switzerland, to be submitted.

[141] J. Ousterhout, M. DeJong, A. Kupries, D. Fellows, K. Lehenbauer, J. Nijtmans, J. Hobbs, G. A. Howlett, D. Porter, K. Kenny, M. Sofer, J. English, and D. Steffen, "Tk Graphical User Interface Toolkit," http://www.tcl.tk, April 2011.

[142] R. Patra, S. Nedevschi, S. Surana, A. Sheth, L. Subramanian, and E. Brewer, "WiLDNet: Design and Implementation of High Performance WiFi Based Long Distance Networks," in *4th USENIX Symposium on Networked Systems Design & Implementation*, Cambridge, MA, USA, April 11-13 2007, pp. 87–100.

[143] PC Engines GmbH, "Wireless Router Application Platform (WRAP)," www.pcengines.ch, 2006. [Online]. Available: www.pcengines.ch

[144] ——, "ALIX system boards," www.pcengines.ch, 2011. [Online]. Available: www.pcengines.ch

[145] T. Perennou, E. Conchon, L. Dairaine, and M. Diaz, "Two-Stage Wireless Network Emulation," in *IFIP World Computer Congress - Workshop on Challenges of Mobility*, Toulouse, France, Aug. 22–27 2004.

[146] C. E. Perkins and P. Bhagwat, "Highly Dynamic Destination-Sequenced Distance-Vector routing (DSDV) for mobile computers," in *Conference on Communications Architectures, Protocols and Applications (SIGCOMM '94)*. London, United Kingdom: ACM, 1994, pp. 234–244. [Online]. Available: http://doi.acm.org/10.1145/190314.190336

[147] B. Pinheiro, V. Nascimento, W. Moreira, and A. Abelém, "Abaré: A Deployment and Management Framework for Wireless Mesh Network," in *IEEE Latin-American Conference on Communications (LATINCOM '09)*, Medellin, Colombia, September 10-11 2009, pp. 1–6.

[148] B. Pinheiro, V. Nascimento, E. Cerqueira, W. Moreira, and A. Abelém, "Abaré: A Coordinated and Autonomous Framework for Deployment and

BIBLIOGRAPHY

Management of Wireless Mesh Networks," in *FMN*, ser. Lecture Notes in Computer Science, S. Zeadally, E. Cerqueira, M. Curado, and M. Leszczuk, Eds., vol. 6157. Springer, 2010, pp. 100–111.

[149] K. N. Ramachandran, E. M. Belding-Royer, and K. C. Almeroth, "DAMON: A Distributed Architecture for Monitoring Multi-Hop Mobile Networks," in *First Annual IEEE Communications Society Conference on Sensor and Ad Hoc Communications and Networks (IEEE SECON 2004)*, Santa Clara, CA, USA, October 4 - 7 2004, pp. 601–609.

[150] K. N. Ramachandran, K. C.Almeroth, and E. M. Belding-Royer, "A Framework for the Management of Large-Scale Wireless Network Testbeds," in *1st Workshop on Wireless Network Measurements (WiNMee 2005)*, Riva del Garda, Trentino, Italy, April 3 2005.

[151] A. Raniwala and T.-c. Chiueh, "Architecture and Algorithms for an IEEE 802.11 -Based Multi-Channel Wireless Mesh Network," in *24th Annual Joint Conference of the IEEE Computer and Communications Societies (INFOCOM 2005).*, vol. 3, Miami, FL, USA, March 2005, pp. 2223 – 2234.

[152] A. Raniwala, K. Gopalan, and T.-c. Chiueh, "Centralized Channel Assignment and Routing Algorithms for Multi-Channel Wireless Mesh Networks," *Mobile Computing and Communications Review*, vol. 8, no. 2, pp. 50–65, 2004.

[153] D. Raychaudhuri, I. Seskar, M. Ott, S. Ganu, K. Ramachandran, H. Kremo, R. Siracusa, H. Liu, and M. Singh, "Overview of the ORBIT Radio Grid Testbed for Evaluation of Next-generation Wireless Network Protocols," in *IEEE Wireless Communications and Networking Conference (WCNC 2005)*, vol. 3, March 2005, pp. 1664 – 1669.

[154] R. Riggio, N. Scalabrino, D. Miorandi, and I. Chlamtac, "JANUS: A Framework for Distributed Management of Wireless Mesh Networks," in *3rd International Conference on Testbeds and Research Infrastructure for the Development of Networks and Communities (TridentCom 2007)*, Orlando, Florida, USA, May 21-23 2007, pp. 1–7.

[155] L. Rizzo, "Dummynet: A Simple Approach to the Evaluation of Network Protocols," *SIGCOMM Computer Communication Review*, vol. 27, no. 1, pp. 31–41, 1997.

[156] S. Rohde, N. Goddemeier, C. Wietfeld, F. Steinicke, K. Hinrichs, T. Ostermann, J. Holsten, and D. Moormann, "AVIGLE: A System of Systems Concept

BIBLIOGRAPHY

for an Avionic Digital Service Platform Based on Micro Unmanned Aerial Vehicles," in *IEEE International Conference on Systems, Man, and Cybernetics (SMC)*. Istanbul, Turkey: IEEE, October 2010.

[157] S. Roy, A. K. Das, R. Vijayakumar, H. M. K. Alazemi, H. Ma, and E. Alotaibi, "Capacity Scaling with Multiple Radios and Multiple Channels in Wireless Mesh Networks," in *First IEEE Workshop on Wireless Mesh Networks (WiMesh)*, Santa Clara, CA, USA, September 26 2005.

[158] Scalable Network Technologies, "The QualNet Network Simulator," http://www.scalable-networks.com/, 2011.

[159] M. Shen and D. Zhao, "TCP Throughput Performance in IEEE 802.11-based Multi-hop Wireless Networks," in *3rd International Conference on Quality of Service in Heterogeneous Wired/Wireless Networks (QShine '06)*. New York, NY, USA: ACM, 2006, p. 23.

[160] M. L. Sichitiu, "Wireless Mesh Networks: Opportunities and Challenges," in *Wireless World Congress*, Palo Alto, California, USA, May 2005.

[161] J. Smart, R. Roebling, V. Zeitlin, R. Dunn, and various developers, "The wxWidgets Project," http://wxwidgets.org/, April 2011.

[162] ——, "wxPython, a Blending of the wxWidgets C++ Class Library with the Python Programming Language," http://www.wxpython.org, April 2011.

[163] Q. Snell, A. Mikler, J. Gustafson, and G. Helmer, "NetPIPE: A Network Protocol Independent Performance Evaluator," in *IASTED International Conference on Intelligent Information Management and Systems*. Washington, D. C., USA: J. S. Wong, June 5-7 1996.

[164] Q. Snell, A. Mikler, J. Gustafson, G. Helmer, D. Turner, and T. Benjegerdes, "NetPIPE: A Network Protocol Independent Performance Evaluator," http://www.scl.ameslab.gov/netpipe/, August 2009.

[165] J. So and N. H. Vaidya, "Multi-Channel MAC for Ad Hoc Networks: Handling Multi Channel Hidden Terminals Using a Single Transceiver," in *5th ACM International Symposium on Mobile ad hoc networking and computing (MobiHoc '04)*. Roppongi Hills, Tokyo, Japan: ACM Press, May 24 - 26 2004, pp. 222–233.

[166] R. Sombrutzki, A. Zubow, M. Kurth, and J.-P. Redlich, "Self-Organization in Community Mesh Networks - The Berlin RoofNet," in *1st Workshop on Operator-Assisted (Wireless Mesh) Community Networks (OpComm)*, Berlin, Germany, September 18-19 2006, pp. 1–11.

BIBLIOGRAPHY

[167] M. R. Souryal, A. Wapf, and N. Moayeri, "Rapidly-Deployable Mesh Network Testbed," in *28th IEEE Conference on Global Telecommunications (GLOBECOM'09)*. Honolulu, Hawaii, USA: IEEE Press, November 30 - December 4 2009, pp. 5536–5541.

[168] B. Staehle, D. Staehle, R. Pries, M. Hirth, P. Dely, and A. Kassler, "Measuring One-Way Delay in Wireless Mesh Networks - An Experimental Investigation," in *4th ACM PM2HW2N Workshop*, Tenerife, Canary Islands, Spain, October 26 - 30 2009.

[169] T. Staub, M. Anwander, K. Baumann, T. Braun, M. Brogle, K. Dolfus, C. Félix, and P. K. Goode, "Connecting Remote Sites to the Wired Backbone by Wireless Mesh Access Networks," in *16th European Wireless Conference, Lucca, Italy*. IEEE Xplore, April 12 - 15 2010, pp. 675 – 682.

[170] T. Staub, M. Anwander, K. Baumann, T. Braun, M. Brogle, P. Dornier, C. Félix, and P. K. Goode, "Wireless Mesh Networks - Connecting Remote Sites," *SWITCH Journal*, pp. 10–12, March 2010.

[171] T. Staub, M. Anwander, M. Brogle, K. Dolfus, T. Braun, K. Baumann, C. Félix, and P. Dornier, "Wireless Mesh Networks for Interconnection of Remote Sites to Fixed Broadband Networks (Feasibility Study)," Universität Bern, Institut für Informatik und angewandte Mathematik, Tech. Rep. IAM-09-007, December 2009.

[172] T. Staub, D. Balsiger, M. Lustenberger, and T. Braun, "Secure Remote Management and Software Distribution for Wireless Mesh Networks," in *7th International Workshop on Applications and Services in Wireless Networks (ASWN 2007)*, Santander, Spain, May 24-26 2007, pp. 47–54.

[173] T. Staub, D. Balsiger, S. Morgenthaler, and T. Braun, "ADAM: Administration and Deployment of Adhoc Mesh networks," http://rvs.unibe.ch/research/software.html, February 2011.

[174] T. Staub, D. Balsiger, S. Morgenthaler, M. Lustenberger, and T. Braun, "ADAM (Administration and Deployment of Adhoc Mesh networks)," in *Demo session for the KuVS Communication Software Award co-located with KiVS'09, Kassel, Germany*, Kassel, Germany, March 6-7 2009.

[175] T. Staub, M. Brogle, K. Baumann, and T. Braun, "Wireless Mesh Networks for Interconnection of Remote Sites to Fixed Broadband Networks," in *Third ERCIM Workshop on eMobility*, University of Twente, Enschede, The Netherlands, May 27 - 28 2009, pp. 97–98.

BIBLIOGRAPHY

[176] T. Staub, R. Gantenbein, and T. Braun, "VirtualMesh: An Emulation Framework for Wireless Mesh Networks in OMNeT++," in *2nd International Workshop on OMNeT++ (OMNeT++ 2009) held in conjuction with the Second International Conference on Simulation Tools and Techniques (SIMUTools 2009)*, Rome, Italy, March 6-7 2009.

[177] ——, "VirtualMesh: An Emulation Framework for Wireless Mesh and Ad-Hoc Networks in OMNeT++," *SIMULATION: Transactions of the Society for Modeling and Simulation International*, first published online July 2010. [Online]. Available: http://sim.sagepub.com/content/early/2010/07/01/0037549710373909.abstract

[178] ——, "VirtualMesh," http://www.iam.unibe.ch/~rvs/research/software.html, 2011.

[179] ——, "VirtualMesh: An Emulation Framework for Wireless Mesh and Ad-Hoc Networks in OMNeT++," *SIMULATION: Transaction of the Society for Modeling and Simulation International, Special Issue: Software Tools, Techniques and Architectures for Computer Simulation*, vol. 87, no. 1-2, pp. 66–81, January 2011, SAGE Print.

[180] T. Staub, S. Morgenthaler, D. Balsiger, P. K. Goode, and T. Braun, "ADAM: Administration and Deployment of Adhoc Mesh networks," in *3rd IEEE Workshop on Hot Topics in Mesh Networking (IEEE HotMESH 2011) affiliated to 12th IEEE Symposium on a World of Wireless, Mobile and Multimedia Networks (WoWMoM 2011)*, Lucca, Italy, June 20 - 24 2011.

[181] T. Staub, S. Ott, and T. Braun, "Experimental Evaluation of Multi-Path Routing in a Wireless Mesh Network Inside a Building," in *5th Workshop on Mobile Ad-Hoc Networks WMAN 2009*, Kassel, Germany, March 5-6 2009.

[182] ——, "Automated Deployment of a Wireless Mesh Communication Infrastructure for an On-site Video-conferencing System (OViS)," in *4th ERCIM Workshop on eMobility co-located with the 8th International Conference on wired/Wireless Internet Communications (WWIC 2010)*, Lulea, Sweden, Lulea University of Technology, May 2010.

[183] M. Stolz, "iPhone/iPad Mesh Deployment Tool for Onsite Video System (OViS)," Bachelor's thesis, University of Bern, Bern, Switzerland, to be submitted.

[184] H. Sun and H. D. Hughes, "Adaptive Multi-path Routing Scheme for QoS Support in Mobile Ad-hoc Networks," in *International Symposium on Per-*

formance Evaluation of Computer and Telecommunication Systems (SPECTS '03), Montreal, Quebec, Canada, July 2003, pp. 408–416.

[185] Z. Tang, , Z. Tang, and J. J. Garcia-Luna-Aceves, "Hop-Reservation Multiple Access (HRMA) for Ad-Hoc Networks," in *IEEE Infocom 1999 (INFOCOM)*, New York, NY, USA, March 21-25 1999, pp. 194–201.

[186] The MadWifi project, "Linux Kernel Drivers for Wireless LAN Devices with Atheros Chipsets," http://madwifi-project.org/, 2009.

[187] The olsr.org Project, "The olsr.org OLSR daemon: an adhoc wireless mesh routing daemon," http://www.olsr.org/, 2009.

[188] A. Tonnesen, "Implementing and Extending the Optimized Link State Routing Protocol," Master's thesis, University of Oslo, Department of Informatics, 2004.

[189] J. Tourrilhes, "Wireless Extensions: A Wireless LAN API for the Linux Operating System," http://www.hpl.hp.com/personal/Jean_Tourrilhes/Linux/Linux.Wireless.Extensions.html, 2010.

[190] ——, "Wireless Tools for Linux," http://www.hpl.hp.com/personal/Jean_Tourrilhes/Linux/Tools.html, 2010.

[191] E. P. J. Tozer, *Broadcast Engineer's, Referencebook*. 200 Wheeler Road, Burlington, MA 01803, USA: Focal Press - an imprint of Elsevier, 2004, vol. 0-2405-1908-6.

[192] A. Tzamaloukas and J. Garcia-Luna-Aceves, "Channel-Hopping Multiple Access," in *IEEE International Conference on Communications (ICC 2000)*, vol. 1, New Orleans, Louisiana, USA, June 18-22 2000, pp. 415 –419.

[193] ——, "Channel Hopping Multiple Access with Packet Trains for Ad Hoc Networks," in *7th International Workshop on Mobile Multimedia Communications (MoMuC 2000)*, Tokyo, Japan, October 2000.

[194] University of Southern California, Information Sciences Institute (ISI), "Ns-2: Network simulator-2," http://www.isi.edu/nsnam/ns/.

[195] A. Varga, "The OMNeT++ Discrete Event Simulation System," in *European Simulation Multiconference (ESM'2001)*, Prague, Czech Republic, June 6-9 2001.

BIBLIOGRAPHY

[196] A. Varga and R. Hornig, "An Overview of the OMNeT++ Simulation Environment," in *1st International Conference on Simulation Tools and Techniques for Communications, Networks and Systems (Simutools '08)*. Marseille, France: ICST (Institute for Computer Sciences, Social-Informatics and Telecommunications Engineering), March 3-7 2008, pp. 1–10.

[197] ——, "OMNeT++: An Extensible, Modular, Component-Based C++ Simulation Library and Framework," http://www.omnetpp.org/, 2011.

[198] E. Weingärtner, "Synchronized Network Emulation," in *Proceedings of the ACM SIGMETRICS Student Thesis Panel*. Annapolis, MD: ACM, 2008.

[199] E. Weingärtner, H. V. Lehn, and K. Wehrle, "Device-Driver Enabled Wireless Network Emulation," in *4th International ICST Conference on Simulation Tools and Techniques SIMUTools 2011*, Barcelona, Spain, March 21-25 2011.

[200] E. Weingärtner, F. Schmidt, T. Heer, and K. Wehrle, "Synchronized Network Emulation: Matching Prototypes with Complex Simulations," *SIGMETRICS Perform. Eval. Rev.*, vol. 36, no. 2, pp. 58–63, 2008.

[201] E. Weingärtner, F. Schmidt, H. vom Lehn, T. Heer, and K. Wehrle, "SliceTime: A Platform for Scalable and Accurate Network Emulation," in *8th USENIX Symposium on Networked Systems Design and Implementation*, Boston, MA, USA, March 30 - April 1 2011.

[202] K. Wessel, M. Swigulski, A. Küpke, and D. Willkomm, "MiXiM (Mixed Simulator): A Simulation Framework for Wireless and Mobile Networks," http://mixim.sourceforge.net/, April 2011.

[203] B. White, J. Lepreau, and S. Guruprasad, "Lowering the Barrier to Wireless and Mobile Experimentation," in *First Workshop on Hot Topics in Networks (HotNets-I)*, Princeton, New Jersey, USA, October 28-29 2002.

[204] B. White, J. Lepreau, L. Stoller, R. Ricci, S. Guruprasad, M. Newbold, M. Hibler, C. Barb, and A. Joglekar, "An Integrated Experimental Environment for Distributed Systems and Networks," in *Fifth Symposium on Operating Systems Design and Implementation*. Boston, MA, USA: USENIX Association, December 9-11 2002, pp. 255–270.

[205] D. Wu and P. Mohapatra, "QuRiNet: A Wide-Area Wireless Mesh Testbed for Research and Experimental Evaluations," in *2nd International Conference on COMmunication Systems and NETworkS (COMSNETS)*, Bangalore, India, January 5 2010.

BIBLIOGRAPHY

[206] D. Wu, D. Gupta, and P. Mohapatra, "QuRiNet: A Wide-Area Wireless Mesh Testbed for Research and Experimental Evaluations," *Ad Hoc Networks*, vol. In Press, Corrected Proof, February 15 2011.

[207] D. Wu, S. Liese, D. Gupta, and P. Mohapatra, "Quail Ridge Wireless Mesh Network: Experiences, Challenges and Findings," University of California, Davis, California, USA, Tech. Rep., 2006.

[208] H. Wu, Q. Luo, P. Zheng, B. He, and L. M. Ni, "Accurate Emulation of Wireless Sensor Networks," in *Network and Parallel Computing (NPC'2004)*, Wuhan, China, October 18-20 2004, pp. 576–583.

[209] S.-L. Wu, C.-Y. Lin, Y.-C. Tseng, and J.-P. Sheu, "A New Multi-Channel MAC Protocol with On-Demand Channel Assignment for Multi-Hop Mobile Ad Hoc Networks," *International Symposium on Parallel Architectures, Algorithms, and Networks*, p. 232, December 7-9 2000.

[210] S.-L. Wu, Y.-C. Tseng, C.-Y. Lin, and J.-P. Sheu, "A multi-channel mac protocol with power control for multi-hop mobile ad hoc networks," *The Computer Journal*, vol. 45, no. 1, pp. 101–110, January 2002.

[211] Z. Ye, S. V. Krishnamurthy, and S. K. Tripathi, "A Framework for Reliable Routing in Mobile Ad Hoc Networks," in *IEEE Infocom 2003 (INFOCOM)*, San Francisco, CA, USA, May 30 - April 3 2003.

[212] Y. Yuan, H. Yang, S. Wong, S. Lu, and W. Arbaugh, "ROMER: Resilient Opportunistic Mesh Routing for Wireless Mesh Networks," in *First IEEE Workshop on Wireless Mesh Networks (WiMesh)*, Santa Clara, CA, USA, September 26 2005.

[213] X. Zeng, R. Bagrodia, and M. Gerla, "GloMoSim: a Library for Parallel Simulation of Large-Scale Wireless Networks," in *12th Workshop on Parallel and Distributed Simulation (PADS '98)*. Banff, Alberta, Canada: IEEE Computer Society, 1998, pp. 154–161. [Online]. Available: http://dx.doi.org/10.1145/278008.278027

[214] Y. Zhang and W. Li, "An Integrated Environment for Testing Mobile Ad-Hoc Networks," in *3rd ACM International Symposium on Mobile Ad Hoc Networking & Computing (MobiHoc '02)*. Lausanne, Switzerland: ACM, June 9-11 2002, pp. 104–111.

[215] A. Zimmermann, M. Güneş, M. Wenig, U. Meis, and J. Ritzerfeld, "How to Study Wireless Mesh Networks: A Hybrid Testbed Approach," in *21st International Conference on Advanced Information Networking and Applications (AINA '07)*, Niagara-Falls, Ontario, Canada, May 21-23 2007, pp. 853–860.

BIBLIOGRAPHY

[216] A. Zimmermann, A. Hannemann, C. Wolff, L. Schulte, and P. Herrmann, "UMIC-Mesh.net - A Hybrid Wireless Mesh Network Testbed," http://www.umic-mesh.net/, April 2011.

i want morebooks!

Buy your books fast and straightforward online - at one of world's fastest growing online book stores! Environmentally sound due to Print-on-Demand technologies.

Buy your books online at
www.get-morebooks.com

Kaufen Sie Ihre Bücher schnell und unkompliziert online – auf einer der am schnellsten wachsenden Buchhandelsplattformen weltweit! Dank Print-On-Demand umwelt- und ressourcenschonend produziert.

Bücher schneller online kaufen
www.morebooks.de

VDM Verlagsservicegesellschaft mbH
Heinrich-Böcking-Str. 6-8   Telefon: +49 681 3720 174   info@vdm-vsg.de
D - 66121 Saarbrücken   Telefax: +49 681 3720 1749   www.vdm-vsg.de

Printed by Books on Demand GmbH, Norderstedt / Germany